Sarah Haley Knowles

Diminished Prosperity

How a Warming Planet Impedes Healthy Families, Communities, and Economies

palgrave
macmillan

Sarah Haley Knowles
Washington, DC, USA

ISBN 978-3-032-07067-8 ISBN 978-3-032-07068-5 (eBook)
https://doi.org/10.1007/978-3-032-07068-5

This Palgrave Macmillan imprint is published by the registered company Springer Nature Switzerland AG.
The registered company address is: Gewerbestrasse 11, 6330 Cham, Switzerland

Diminished Prosperity

"Climate change and reproductive injustice are not parallel crises but deeply intertwined forces that erode the foundations of family, community, and economic prosperity, as Dr. Sarah Haley Knowles shows in this essential analysis. Practitioners, policymakers, and scholars committed to sustainable equity and systemic justice will benefit from this set of actionable frameworks to advance sustainable prosperity and family well-being."

—Dr. Elizabeth Stearns, Professor of Sociology, Director of the Public Policy program, *University of North Carolina at Charlotte*

"Dr. Sarah Haley Knowles demonstrates the inseparable ties between climate action and reproductive justice. Reading this book will sharpen policymakers' and business leaders' understanding of the subtle economic consequences of environmental choices."

—Dr. Stephanie Moller, Professor of Sociology, *University of North Carolina at Charlotte*

"Dr. Sarah Haley Knowles presents a timely, values-driven framework that reclaims market logic for the public good—reminding us that any sustainable future must be rooted in community, health, and the right to self-determination. Her model of Engage, Evolve, Enhance, and Expand offers a pragmatic and principled path forward in an era of compounding crisis."

—Dr. Robert Pinka, Director of Impact & Policy, *Council for Children's Rights*

"This book offers a comprehensive review and an important new analysis of how climate change intensifies existing reproductive health disparities, particularly for women of color and low-income communities. It makes a timely and valuable contribution to the growing literature on environmental justice and public health."

—Dr. Stephanie Potochnick, Professor of Sociology, Director of the Center for Migration & Diaspora Studies, *University of North Carolina at Charlotte*

Preface: Public Perceptions on Climate Change

In November 2023, a summary of the findings of the Intergovernmental Panel on Climate Change (IPCC) Sixth Assessment Report was published, highlighting the state of climate change, its widespread impacts and risks, and proposed solutions for climate mitigation and adaptation. With a target audience of policymakers, the report outlines the interdependency of climate, ecosystem services, and human societies, along with the relationship between a changing climate, ecosystem health, human well-being, and sustainable development. Among the findings is unequivocal evidence that the past two decades saw global surface temperatures reach 1.1 degrees Celsius above temperatures previously recorded during the preceding 150 years. The report concludes with high confidence that global greenhouse gas emissions are caused by human activity, particularly through the emitting of greenhouse gases caused by unsustainable energy use, land use, industrial production, and consumer lifestyle and consumption behavior (Lee et al., 2023).

Anthropogenic climate change, also known as human-induced climate change, refers to the long-term changes in the Earth's climate primarily caused by human activities. In 2021, a study explored the main causes of climate change, including over-exploitation of natural resources, agricultural practices, landfill use, deforestation, pollution, and natural cyclical changes, and concluded that five of the six main causes are directly tied to human activity (Panda & Maity). The scientific consensus is that human activities are the main cause of the observed warming since the mid-twentieth century, isolating the burning of fossil fuels as the primary source of emissions trapping heat in the atmosphere that causes the planet to warm. Unabated global warming has led to a range of adverse impacts on the planet, including sea level rise, changes in precipitation patterns, more frequent and severe extreme weather events, and biodiversity loss. Yet, even with an overwhelming scientific consensus on the cause and impact of climate change, most Americans disconnect their day-to-day activities from the consequences of a warming planet, content with associating adverse climate impacts with something tragic that happens to other people and in foreign communities in faraway places.

When people think about climate change, they often visualize extreme hurricanes, typhoon-style rainstorms, crippling heatwaves, or abstract impacts to communities located worlds away. It is not uncommon for extreme weather to be associated with apocalyptic blockbusters, media-driven fearmongering, and complicated, hard-to-understand science talk, which may render it difficult for the average American to see the link between climate change and their everyday life or be motivated to look for it. Of those who may make the connection between extreme weather conditions and adverse impacts to food, housing, and healthcare, they may also dismiss the unsettling reality as something happening to people elsewhere, not as an imminent danger to them or their neighbors, who are often preoccupied by their busy lives, vibrant social media feeds, and inbox tickers. This may partly be because they do not have an in-depth understanding of the relationship between climate and local economies, and partly due to fear or a refusal, likely a subconscious one, to accept the harsh realities of climate change, as jarring and overwhelming as they are. Whatever the cause, my research explores the possibility that the same consumer-driven, corporate lifestyles fueled by jam-packed schedules, bigger-is-better mantras, and instant gratification, single-use, convenience-oriented solutions that are driving climate change, may also be keeping most Americans too busy to notice the slow, but steady, changing climate around them, resulting in most Americans being ill-prepared for the adverse impacts of climate change.

Ironically, however, the development, promotion, and protection of strong, healthy families; safe, vibrant communities; and flourishing, resilient economies is often listed at the top of both constituents' priorities and policymakers' agendas, regardless of party affiliation. Core family values, livable neighborhoods, and growing businesses are top of mind for most voters. Protecting the ability to realize the "American dream" is lauded by politicians on both sides of the aisle as what they fight for. Yet, the same policymakers who pledge to protect the right and ability of all Americans to secure safe housing, nutritious food, and clean bills of health are also those who often fail to tackle the root causes of homelessness, food insecurity, and poor health outcomes. Further, while the inability to afford adequate housing, food, and healthcare is repeatedly credited as a reason people choose not to have children or to have less children, rarely do policymakers link family planning decisions to an increasingly significant underlying cause of economic instability: climate change, or the change in the planet's average temperature primarily caused by human activities.

A 2014 *Pew Research* poll found climate change ranked last on a list of 21 issues that the government should prioritize for action ahead of the 2012 Presidential Election. Two elections later, a similar survey explored voter priorities ahead of the 2020 election and found that climate change rose to number 13 on a list of 30 priority issues, with 60% of voters stating that climate change is an important issue when deciding who to vote for, an increase of 14 points from the 2016 election. However, only 4% of Americans polled identified climate change as the most important issue. In addition, while most registered voters believe that corporations, policymakers, and communities should be doing more to address climate change, only 2% are currently participating in climate action, and only 13% claim they have reached out to an elected official to encourage climate action in the last 12 months (Leiserowitz et al., 2020a, b). Ahead of the 2024 election, *Pew Research* found that 69% of Americans believe climate change is an important issue, with 36% wanting their next President to address it as a top priority.

While recent studies evidence that Americans are concerned about climate change, most voters continue to place climate action toward the bottom of their priority list. Accordingly, a 2019 poll from the *Yale Program on Climate Change Communication* found that 45% of Americans believe that people in the United States are currently being harmed by global warming, with 43% stating they believe they will be harmed by climate change, 49% believing their family will be harmed, and 56% believing climate change poses a risk to their community. Yet, 33% of Americans state that climate change is not personally important to them, with 59% indicating they do not discuss climate change with their family or friends and only 40% perceiving an expectation for individuals and American households to act on climate change. While most Americans think global warming is affecting weather in the United States, causing concern about harm from extreme weather events, approximately half of Americans believe climate change will be solved with the use of new technologies without requiring changes to lifestyles or consumer behavior (Leiserowitz et al., 2020a, b).

With over half of Americans claiming to understand the impacts of climate change on their families and communities, why does acting on climate change continue to take a back seat to other voter issues, such as inflation, employment, community safety, and healthcare? A 2023 report by the United States Department of the Treasury (USDT) illustrates how American households will experience significant financial strain due to climate change, referencing increases in groceries, utilities, and housing costs, including $56.92 billion of climate-induced property damages impacting one in ten American households for the year 2021 alone (USDT, 2023). However, are heads of households accounting for climate-related expenses? And if not, what are the consequences of not being prepared?

Global warming and the associated impacts of climate change can be unpredictable. Without proper planning and social and economic safety nets, most households will suffer from the adverse impacts of a changing climate. What is less clear, however, is how climate change influences individual concepts of freedom, or the ability to freely choose a travel plan or career path; a couple's or family's realization of full autonomy over food choices, living arrangements, and fertility desires; and society's understanding of prosperity, feminism, and justice within the constructs of family planning. The inequitable impacts of a changing climate make it not only harder to address the root causes of global warming but also create new barriers for maternal health outcomes and the attainment of a gender-equal world. Deeply embedded in the advent of extreme weather events, higher temperatures, prolonged droughts, and failing crops, is the inability for individuals, families, and communities to choose their reproductive destinies. Although often ignored, misunderstood, or exploited, addressing the link between climate change and reproductive freedom is essential for the continued protection and fulfillment of full bodily autonomy, gender equality, and social justice.

While an awareness of the comprehensive relationship between climate change and feminism is limited, the link between the two is not an entirely new concept. In 1993, the iconic "*Ecofeminism: Women, Animals, Nature*," edited by Greta Gaard, was published. In it, a series of essays rejected the nature and culture duality derived from patriarchy while exploring how the oppression of nature mirrors the oppression of women (Larocque, 1993). More recently, in 2019, Ireland's former and first female President, Mary Robinson, published "*Climate Justice: A man-made problem with a feminist solution*" to shed light on how the climate crisis is fundamentally a crisis of humanity by highlighting how systemic issues like sexism, racism, and classism both create and sustain global warming. Indeed, climate change is not just about greenhouse gas emissions; it is about the violation of human rights through rampant food shortages, unemployment, infectious disease, and homelessness, all of which ultimately impact the ability of families, communities, and economies to flourish, thrive, and survive. Therefore, my research explores how climate change is born of an oppressive patriarchal system spun by threads of white supremacy, extractive capitalism, and unequal distribution of power and consequence.

In 2021, researchers Zalak Desai and Zhanghat Ying examined 35 academic articles that explored the link between climate and reproductive health and concluded that reproductive health outcomes are highly vulnerable to the impacts of climate change, with societal, cultural, and economic factors determining the level of correlation. Their findings include increased climate vulnerability for women, due to extreme weather events adversely impacting the agricultural industry where women are primarily employed, and, in turn, increasing the risk of nutritional deficiencies, food insecurity, and vector-borne diseases, with pregnant women more likely to experience exhaustion and hypertension leading to an increased risk of miscarriage and stillbirth (Desai & Zhang, 2021). Still, while many studies show that climate change has a significant impact on the health and well-being of individuals and communities around the world, especially women, few explore how a changing climate restricts an individual's bodily autonomy through the lens of reproductive justice, particularly how a changing climate impacts American household budgets and, as a result, restricts the ability to have and parent healthy children in safe and healthy environments.

When people think about reproductive freedom, most picture the reproductive rights movement focused on protecting universal access to contraceptives and abortion. However, the reproductive rights framework provides us with a roadmap for addressing only one aspect of reproductive freedom: protecting the right to not have a child. Reproductive justice, on the other hand, provides us with the framework for addressing all components of reproductive freedom: protecting the right to not have a child, the right to have a child, and the right to parent a child in a safe and healthy environment. First coined by the *Women of African Descent for*

Reproductive Justice, a group of Black women who gathered in Chicago during the summer of 1994 to craft the basis for addressing reproductive oppression, reproductive justice was born out of the understanding of social justice and the failures of the reproductive rights movement. Women of color, who for centuries suffered the worst forms of reproductive violence, were largely excluded from the white women-led pro-choice campaigns and desired a new framework to protect themselves and their communities from reproductive oppression, or the "control and exploitation of women and girls through their bodies, sexuality, and reproduction as a strategic pathway to regulating entire populations that is implemented by families, communities, institutions, and society" (Venkatraman, 2020).

The reproductive justice framework was created to advocate for freedom "based in the human right to make personal decisions about one's life, and the obligation of government and society to ensure that the conditions are suitable for implementing one's decisions" and to govern the ability to freely choose to not have a child, to have a child, and to parent children in a safe and healthy environment (Ross, 2017). Therefore, as climate change increases health risks and barriers to comprehensive healthcare, which infringes upon the right to freely choose to have a child or not to have a child, and decreases the availability of affordable, adequate housing and nutritious food, which limits the ability to raise children in safe and healthy environments, climate change also worsens maternal health outcomes, exacerbates reproductive health inequities, and restricts the ability to realize fertility desires.

To help address the gaps in the literature, I conducted a new study in 2024 to examine the relationship between the adverse impacts of climate change, measured through soil moisture, increased daily temperatures, and increased incidents of extreme weather, and the cost of living, represented by increases in food prices, housing costs, and healthcare expenditures. The relationship between climate change and living costs is then measured against trends in birth rates to better understand how a changing climate interacts with the realization of fertility desires. As discussed throughout this book, findings show a positive correlation between the consequences of a changing climate and increased costs of living, and a negative correlation between increased living costs and birth rates. Yet, while the insights gained from this study are powerful, the findings lend more questions than answers to an increasingly urgent predicament faced by humanity: is climate change to blame for recent trends in the rising cost of living and the declining rate of births, and, if so, what do we do about it?

An increased understanding of the relationship between climate change and reproductive injustice will contribute to society in several ways, including (1) deepening the understanding of the relationship between climate change and food prices, affordable housing, and access to healthcare; (2) exploring the social implications of rising food prices, rapid urbanization, higher temperatures, and increased infrastructure collapse; (3) understanding how these impacts create additional barriers to family planning decisions; (4) designing and testing new models of prosperity as potential climate change mitigants; and (5) assisting with agenda-setting of public policy priorities to address adverse impacts of climate change and promote healthy families, communities, and economies. With increased knowledge of how a changing climate interacts with fertility desires, living costs, and market viability, policymakers and industry leaders can make more informed, effective, and strategic decisions to promote, protect, and advance prosperity among families, communities, and economies.

Organized into four parts, this book examines the relationship between climate change, reproductive injustice, and the health of families, communities, and economies. The first part builds a foundation for examining how climate and prosperity interact by summarizing the climate change science literature that addresses the health and social harm disparities for low-income communities, as compared to other communities, with a particular emphasis on disproportionate impacts to low-income women of color, and the doctrines of full bodily autonomy, reproductive justice, and green criminology to establish the connections between reproductive injustice and the adverse impacts of a changing climate. The second part builds upon this foundation by recalling the historical legacy of settler colonialism, especially the exploitation of Indigenous women and their ongoing oppression, and examining how corporate expansion, consumerism, and white feminism create and maintain corporate colonialism, drive climate change, and ultimately contribute to the colonization of low-income communities and communities of color. Part III explores how a modern society measures prosperity, along with what the role of family, community, and economy are within the constructs of a prosperous society.

Lastly, in Part IV, I introduce a new model for measuring prosperity based on a four-part test—engage, evolve, enhance, and expand—which promotes, protects, and advances the health of families, communities, and economies through sustainable, responsible, and economic principles, means, and indicators of success. Through my research, I hope to strengthen and expand our collective understanding of both the drivers and the lasting consequences of climate change so that we can put forth effective policies and solutions to address the root causes of a changing climate, while also protecting the health and prosperity of society, the planet, and the economy.

Washington, DC, USA Sarah Haley Knowles

References

Desai, Z., & Zhang, Y. (2021). Climate change and women's health: A scoping review. *Geohealth, 5*(9), e2021GH000386.

Larocque, M. (1993). Ecofeminism: Women, animals, nature. *Canadian Woman Studies/les cahiers de la femme.*

Lee, H., et al. (2023). *IPCC, 2023: Climate change 2023: Synthesis report, summary for policymakers.* Contribution of Working Groups I, II and III to the Sixth Assessment Report of the Intergovernmental Panel on Climate Change [Core Writing Team, H. Lee and J. Romero (Eds.)]. IPCC.

Leiserowitz, A., et al. (2020a). *Climate change in the American mind: November 2019.*

Leiserowitz, A., et al. (2020b, April). *Politics and global warming.*

Ross, L. J. (2017). Conceptualizing reproductive justice theory: A manifesto for activism. In *Radical reproductive justice: Foundations, theory, practice, critique* (pp. 170–232).

United States Department of the Treasury. (2023). *The impact of climate change on American household finances.*

Venkatraman, R. (2020). *A new vision for advancing our movement for reproductive health, reproductive rights, and reproductive justice (2005), by Asian communities for reproductive justice (ACRJ).* Embryo Project Encyclopedia.

Acknowledgments

I express my sincerest gratitude to those who helped, supported, and guided me on this journey:

To my academic advisor, Dr. Stephanie Moller, for her invaluable advice, continuous guidance, support, encouragement, resilience, and patience during my PhD research study.

To the Economics, Sociology, Political Science, Public Health, and Gender Studies faculty of the University of North Carolina at Charlotte, namely Dr. Mike Dulin, Dr. Elizabeth Stearns, Dr. Redwan Bin Abdul Baten, Dr. Bruce Arrigo, Dr. Andrea Pitts, Dr. Stephanie Potochnick, Dr. Robert Pinka, Dr. Elizabeth Paquette, and many others, for their guidance, credible challenges, and instrumental teachings throughout this process.

To my most transformational teachers, namely Tina, Mars, Maya, Hannah, Ash, Christina, Christine, Taylor, Jess, Nikki, Justina, Angela, and all those at SisterSong, Carolina Reproductive Action Network, Climate Reality, Carolina Abortion Fund, Planned Parenthood, and the many other community organizations who fill in the gaps where systems, institutions, and societies fall short.

To my partners in progress, namely Beth, Miguel, Jonathan, Eugene, Okeatta, Mary, Rachael, Jennifer, Ramsay, CaSondra, Tim, Jennivine, Ricardo, Talya, Keri, and many others who have not only supported me on this journey but who every day challenge the status quo in Corporate America and demand that we do better.

To Lauren, Kadena, Ilea, Paige, Ashleigh, Traci, and Mary, who have always seen me as the best version of myself, and whose unwavering support lined my path on this journey.

And, finally, to Uncle Joe, the first Dr. Haley, who demonstrated what excellence looks like and inspired me to believe in what was possible; to Uncle Pat, who taught me from a young age the importance of and value in speaking truth to power even when my voice shakes; and to Aunt Jean, who showed me what it means to know, love, and care for thyself unconditionally.

About This Book

Over the course of four generations, the fight for gender equality has been built upon the fight for reproductive rights, namely the right for a woman to have an abortion. The literature is ripe with discussions of the reproductive rights framework and studies demonstrating the positive correlation between women's socioeconomic status and access to abortion (e.g., Bernstein, 2018; Foster et al., 2022; Everett & Taylor, 2024), rightfully acknowledging the criticality of full bodily autonomy in achieving full prosperity. Yet, largely absent from this discussion (and the literature) is the importance of protecting the right to have a child and the right to raise a child in a safe and healthy environment, along with the understanding that full bodily autonomy cannot be achieved without the protection of both. Further, climate change as an obstacle to realizing fertility desires and, therefore, full prosperity is not well-researched or understood within this context and within the literature.

This book explores the relationship between climate change, reproductive injustice, and the prosperity of families, communities, and economies through a comprehensive literature review, a new theoretical framework, and the findings of a new study that examines the relationship between the adverse impacts of a changing climate and the cost of living, represented by increases in food prices, housing costs, and healthcare expenditures, and the associated impact on declining birth rates across the United States. The study results support the hypothesis that indicators of a changing climate positively correlate with increased costs of living and decreased birth rates, paving the way for the introduction of a new prosperity model based on a four-part test to promote, protect, and advance the health and long-term viability of families, communities, and economies through sustainable and responsible economic principles, means, and indicators of success.

Part I establishes the foundation for a new theoretical framework by first examining how climate change, reproductive justice, and societal prosperity interact with a review of the climate change science literature, including the unequal consequences of climate change based on the health and social harm disparities for women of color and low-income communities, and the need for the reclaiming of full bodily autonomy given who is most impacted. Part II builds upon this foundation by recalling the historical legacy of settler colonialism, especially the exploitation of indigenous women; examining how the history and continuation of consumerism, white feminist ideology, and corporate expansion create and maintain the oppression of nonwhite women in the name of economic and societal progress; and calling for a reconstruction of feminism and corporate success metrics toward a new model that promotes, measures, and protects prosperity for all. Part III completes the theoretical framework by exploring how a modern society measures prosperity through a review of classical economic theory and historical policy approaches to promoting healthy families and communities. Woven throughout all three parts is a discussion of how the consequences of a changing climate create and maintain barriers to full bodily autonomy viewed through the lens of an individual's ability to freely choose to have a healthy child and to parent in safe and healthy environments.

The results of a new study that measures the relationship between outputs of climate change and components of reproductive justice are summarized and discussed. Using state-level panel data over a 25-year period and four fixed effects linear regression models, the relationships between (1) soil moisture and food expenditures, (2) extreme weather events and housing expenditures, (3) surface temperatures and healthcare expenditures, and (4) the cost of living and birth rates are examined. Findings show a positive correlation between consequences of a changing climate and increased costs of living and a negative correlation between increased living costs and birth rates. The findings support the need for a new model of prosperity, which is introduced in Part IV based on four market approaches—engage, evolve, enhance, and expand—and the preconditions, benefits, and goals attached to each one. Implications and considerations for private industry and public policy are discussed, along with the need for additional research to better understand the new model's application to and effectiveness in predicting, determining, and protecting societal prosperity at both micro and macro levels. It will be of interest to researchers in gender studies, feminist theory, sustainable economics, environmental sociology, sustainability studies, and social policy.

References

Bernstein, A. (2018). *Factors associated with qualifying for abortion funds: An exploratory analysis.*

Everett, B. G., & Taylor, C. J. (2024). Abortion and women's future socioeconomic attainment. *American Sociological Review, 89*(6), 1044–1074.

Foster, D. G., et al. (2022). Socioeconomic outcomes of women who receive and women who are denied wanted abortions in the United States. *American Journal of Public Health, 112*(9), 1290–1296.

Contents

About the Author

Sarah Haley Knowles is an ESG strategist and social scientist with over 15 years of experience coordinating the development, testing, and enabling of global sustainability and social impact initiatives for Fortune 50 companies. She has a passion for evolving corporate strategy and public policy to assign value to sustainable, equitable, and inclusive economic activities measured by both financial and nonfinancial performance at the triple bottom line: people, planet, and profit. Her research explores how environmental and social issues intersect and compound to create complex adverse impacts for families, communities, and economies through the lens of environmental, social, and economic prosperity and long-term viability. Sarah holds a Master of Business Administration from Drake University, a Doctor of Philosophy in Public Policy from the University of North Carolina at Charlotte, and numerous certifications in the field of ESG and Sustainability. She resides in Northern Virginia, where she enjoys gardening, beekeeping, hiking, teaching, and spending time with her foster children, English bulldogs, and African gray parrot. A staunch advocate for business as a force for good, she continues to help companies evolve traditional shareholder-primacy models toward a multistakeholder approach in corporate engagement, governance, and decision-making.

List of Figures

List of Tables

PART I

Ethical Considerations at the Intersection of Climate Change and Reproductive Injustice: Understanding Who Is Most Impacted by Climate Change

CHAPTER 1

The Gendered Experience of Climate Change

Climate change does not impact us all equally. The impacts of extreme weather and climate events are influenced by societal structures. Socioeconomic power, cultural norms, and traditional roles make some people more vulnerable to climate change than others. Therefore, the consequences of rising sea levels, increased temperatures, drier conditions, and more extreme storms are disproportionately felt by low-income communities and women, particularly women of color. As climate change increases, the harm caused to women and young girls infringes upon their bodily autonomy. Yet, long absent from the climate debate is the issue of full bodily autonomy and reproductive justice.

The first part of this book discusses who is most impacted by climate change and highlights several ethical concerns that warrant public policy attention at the intersection of full bodily autonomy and the transition to a low carbon economy. The first four chapters explore how barriers to nutritious food, safe housing, adequate healthcare, or anything that obstructs the right to freely choose to have a child, to not have a child, or to raise a child in a safe and healthy environment diminishes bodily autonomy and, therefore, creates reproductive injustices. In addition to summarizing the climate change science literature addressing the health and social harm disparities for low-income communities, particularly women of color, in the second chapter, I discuss the findings of my recent study on the relationship between climate change and the cost of living. While

S. Haley Knowles, *Diminished Prosperity*,
https://doi.org/10.1007/978-3-032-07068-5_1

full bodily autonomy is vital for equitable communities, reproductive justice is often ignored by policymakers seeking to solve and mitigate climate change, further exacerbating the injustice.

In 2011, researcher Seema Arora-Jonsson pointed out, “It is clear that more context specific case studies are needed to understand the linkages of gender and climate change—comparative case studies that examine not only relationships and adaptations on the ground but also ask new questions from different vantage points” (Arora-Jonsson, 2011, p. 749). Responding to a collection of studies on the relationship between climate and gender, Arora-Jonsson acknowledged the gaps in the research. Since then, additional studies continue to explore the impacts of climate change as experienced through social structures and cultural power dynamics. Historically oppressed populations experience climate change more intensely than those with privilege and power. Gender, along with poverty and education level, is an indicator of vulnerability to climate change (Alston, 2013). As such, the literature shows that women are impacted by climate change more than men, and low-income women and women of color are impacted the most.

Globally, women and young girls have lower socio-economic power than men and experience poverty at a much higher rate. The same cultural, historical, political, and economic factors that influence the cycle of poverty disproportionately impacting women are also responsible for the inequitable distribution of climate change impacts. A 2015 study conducted on the impacts of climate events on women in Bangladesh, largely considered one of the countries with the worst climate impacts today, concluded that “the distribution of five livelihood capitals (human, natural, financial, social, and physical) of women are heavily influenced by several climatic events, such as cyclones” (Asaduzzaman, 2015, p. 5). In communities on the frontlines of climate change, the roles women and young girls fulfill are decided by the cultural and religious norms of the region. Women and girls are likely to be caregivers and responsible for gathering water and fuel for fires for their families. Women are also responsible for approximately 50% to 80% of the global food-producing workforce but only own around 20% of the land (Pachauri & Meyer, 2014). This leaves them without a safety net when disasters strike, which often destroy crops, housing, and other necessities, and makes them dependent on others for their livelihood. In addition, women are usually responsible for caring for children and aging parents, which restricts their ability to evacuate when extreme weather occurs. As a result of the roles women traditionally fill, such as primary caregivers and providers of food or fuel, they, along with young girls, are more vulnerable when climate events occur, such as hurricanes, flooding, or droughts (Alston, 2013).

Displacement, Poverty, and Housing

Climate change directly impacts women and young girls whose livelihoods depend upon water, wood, and food resources. With increased temperatures, stronger storms, flooding, drought, and crop depletion; crop yields, water, and wood sources are becoming more scarce and unreliable due to a changing climate. This results in women and young girls who fulfill these roles having to walk farther every day to reach the resources or work harder to plant and harvest additional crops (Pachauri & Meyer, 2014). In addition to the health impacts of increased labor, these added hours and increased work limit the time women and young girls have to perform other tasks, such as attend school or acquire new skills. These intense workloads often result in young girls leaving school to assist their family with chores (Lane & McNaught, 2009). This perpetuates a cycle of poverty not only in communities on the frontlines of climate change but in communities all around the world where women have less socioeconomic power and tend to take on more traditional roles.

One of the biggest consequences of climate change is an increase in extreme weather events. The ability to prepare for, survive, and recover from natural disasters is contingent on many factors, including gender. The same cultural, historical, political, and economic factors that lower socioeconomic power and increase the rate of poverty for women and young girls also create disparities in casualties from natural disasters. For example, in frontline communities, women are more likely to be responsible for disaster preparedness of the household, including food and water preparation, organizing household belongings, notifying family members, and caring for young children and elderly parents. As a result, women have more work to do before and after a disaster and less time to evacuate (Alston, 2013). Shockingly, a 2005 study on the aftermath of a tsunami in Indonesia, Sri Lanka, and India found the rate of men who survived the storm outnumbered surviving women by four to one. The study concluded, "the reasons for this are similar across [all the] countries: women died because they stayed behind to look for their children and other relatives. Women in these areas often swim or climb trees, which meant that they couldn't escape" (MacDonald, 2005, p. 474).

Further, a study conducted in 2007 looked at natural disasters in 141 countries over a 20-year period, along with casualties based on gender and the level of socioeconomic power women typically possess within the associated culture. The study found that women represent the majority of those most negatively impacted by climate disasters. Specifically, women are approximately 14 times more likely to die in a disaster than men. As a result, women's life expectancy is lowered from disasters at a much higher rate than men's, more women are killed from disasters than men, and women are killed at a much younger age during disasters than men are. In addition, the relationship between the intensity of the climate disaster and women's rights and the impact on the gender gap in life expectancy is well documented. As the strength of a disaster increases, the gender gap in life expectancy also increases. Importantly, these disparities are reduced in countries where women have increased socioeconomic power (Neumayer & Plümper, 2007).

For women and young girls able to evacuate before a climate event occurs or fortunate to survive a natural disaster, returning home and recovering from a disaster can be very challenging. With lower socioeconomic status, women, particularly women of color and single-mother households, experience poverty at a higher rate and, therefore, suffer more long-term impacts from climate events than men do. This complicates and extends recovery efforts for these households, even in the presence of disaster recovery aid. For example, a 2007 study found that women are more likely to be excluded from disaster aid distribution due to their roles and needs being undervalued and often ignored by aid organizations and community workers (Dominelli, 2013).

Often, the barriers to returning home following a disaster go beyond just housing. For example, rebuilding efforts, particularly construction jobs, which are traditionally male-dominated fields, frequently are the main source of employment following climate disasters. For example, a 2008 study found that in the aftermath of Hurricane Katrina "the odds of holding no job or a worse job were 1.97 times greater for women than for men and 0.83 times lower at higher income" (Zottarelli, 2008, p. 603). In August 2010, the Institute for Women's Policy Research published a fact sheet explaining why women were disproportionately impacted by Hurricane Katrina. Factors include housing status, access to transportation, lack of childcare options, mobility issues, and age, among others (IWPR, 2010). A 2024 study found that 158 million women and girls are at risk of being pushed into poverty by 2050 due to the consequences of climate change, which is 16 million more than the number of men and boys predicted to be impacted by the same events (UN Women, 2025).

Accordingly, a study published in 2015 found that Black women, especially single mothers, were most impacted by Hurricane Katrina. The study analyzed interviews with 184 low-income Black women who were living in one of the four largest public housing projects when Hurricane Katrina made landfall in 2005. More than half the low-income households located in impacted areas were headed by single Black mothers. Respondents "were less concerned about getting back to traditional public housing, particularly as that option disappeared, and were more preoccupied with getting a job or another source of income to pay their expenses, including the added costs such as health and childcare, transportation, higher rents, and utilities" (Henrici & Shaw, 2015, p. 25). One of the major findings in the study was the erosion of community networks and safety nets following the disaster and the impact on single-mother households. Because single-mother households are reliant on interdependent community networks for everyday needs, the direct impacts of the storm on infrastructure, jobs, public transportation, and housing prevented most Black women living in public housing from being able to return home (Henrici & Shaw, 2015).

As a result of mass displacement, climate change is a primary driver of migration, including both migration within countries of residency and immigration to other countries. As previously discussed, many communities in developing countries, often those on the frontlines of climate change, are reliant upon food production for their livelihood. These same communities are experiencing declines in crop yields and food production due to climate change. Climate scenarios predict this trend will increase as global warming increases, along with water scarcity, drought, and erratic weather patterns. As crop yields decline, saltwater intrusion increases due to sea level rise, and extreme weather continues to impact coasts, crops will fail. Those most impacted by these trends are low-income food producers who do not own land and who have little to no socioeconomic power, mainly women (Pachauri & Meyer, 2014). As a result, migration has increased exponentially, particularly in developing countries such as Central America and the Middle East, as bankrupt farmers and displaced populations seek refuge and new opportunities. The literature predicts the flows of migrants will increase due to crop failure and the inability of vulnerable populations to adapt to a changing climate (Klepp, 2017).

As crop yields continue to decline and displacements increase, rural to urban migration is on the rise. As a result, rapid urbanization, or sharp increases in population in urban areas, is occurring which increases competition for housing and, in return, increases housing prices (Tacoli et al., 2015). In addition, as evidenced by Henrici and Shaw's study (2015), as climate disasters increase, women and single-mother households displaced by such events are forced to resettle elsewhere. For those without the ability to move closer to family, the process of starting over in a new city without a support system increases the vulnerabilities created by displacement and makes recovery even harder. Urban areas with a lack of affordable housing and without strong community safety networks are associated with high incidences of housing insecurity and homeless residents. Based on these studies, the literature shows a relationship between climate change and homelessness due to displacement and increased housing costs. Those most impacted by these increased costs are women and single-mother households (Henrici & Shaw, 2015).

Increased Food Insecurity and Malnutrition

Along with migration and housing security, food security is also impacted by declines in farm yields. Climate change is expected to increase world food prices by 30% to 50% in the next few decades and increase food prices by 84% by 2050 (Pachauri & Meyer, 2014). Social issues, such as food insecurity and affordable housing, increase and become more complicated in the presence of a changing climate. Accordingly, climate change is considered the greatest threat to food security for vulnerable populations (Wang, 2010).

A study in 2013 found that "changes in food price may result in consumers changing their diets" (Bradbear & Friel, 2013, p. 64). If the price of healthy food increases, less healthy food may be substituted by those who cannot afford to pay the higher price. This will impact vulnerable populations the most, particularly those living in food deserts (Bradbear & Friel, 2013). Further, the adverse effects of climate change on food and nutrition security are not evenly distributed among men and women. For example, in sub-Saharan Africa, climate-induced food and nutritional insecurity disproportionately impact more women than men due to socially defined gender roles (Awiti, 2022).

The lack of access to nutritional food can exacerbate preexisting medical conditions, which is often tied to increased health issues, both physically and mentally (Bloem et al., 2010). Women are especially sensitive to climate-driven food insecurity and malnutrition due to increased needs during menstruation, childbirth, and breastfeeding (Sorensen et al., 2018). In 2021, researchers concluded that climate-driven malnutrition among women can be traced back to women skipping meals to feed other household members when resources are restrained, along with traveling further to find water and fuel for their household. As a result, increased food insecurity and malnutrition are most at risk in historically oppressed communities and those without the means to adequately adapt to climate change—which is most often women.

The literature shows that access to safe housing and nutritious food is vital for good overall health, and a lack of access to housing and nutritional food exacerbates preexisting conditions, which is tied to increased health issues, both physically and mentally. The World Health Organization published a robust framework on the social determinants of health in 2008, which evidences that "socially determined factors that influence health outcomes include rapid and often adverse social, economic and demographic changes that affect working conditions, learning environments, family patterns, as well as the culture and social fabric of communities" (WHO, 2008, p. 9). The connection between social structures and health outcomes is more pronounced for women because "social determinants of health contribute significantly to premature death and diseases particularly among vulnerable groups such as women" (WHO, 2008, p. 8).

Access to Healthcare and Maternal Health Outcomes

As climate change increases infectious diseases, allergies, and heat-related illnesses, women are most impacted, particularly pregnant women. A study conducted in 2018 concluded that pregnant women, or those trying to become pregnant, are one of the most at-risk groups for the Zika virus and that "temperature is a strong driver of vector-borne disease transmission" (Tesla et al., 2018, p. 5). The study explored the relationship between temperatures and the ability of mosquitoes, which transmit disease, to live longer and to travel to warming regions previously unsuitable for insects to survive, and concluded that climate change increases the transmission of Zika and, therefore, the risk of infection for pregnant women (Tesla et al., 2018).

As climate change increases health problems, it also creates additional risks for healthcare facilities due to extreme weather events. Extreme weather events can impact access to healthcare facilities and the delivery of care. This can lead to disruptions in care and poorer health outcomes for individuals who have limited options for care (Paterson et al., 2014). Extreme weather events can impact the delivery of services at hospitals in the days and weeks following an extreme weather event, due to collapsed infrastructure and scarcity of resources and staff. Communities on the frontlines of climate change already face many challenges in accessing healthcare. In addition, maternal health outcomes tend to be lower in these same communities. Higher occurrences of extreme weather and climate events increase the vulnerabilities already present within healthcare systems, particularly for maternal health outcomes for women and young girls.

Sexual and Intimate Partner Violence

Climate change can also increase sexual and domestic violence for women and young girls as women experiencing poverty, isolation, and life disruption tend to be more vulnerable to sexual exploitation and domestic violence. Research shows that "women and girls are at a higher risk of physical, sexual, and domestic violence in the aftermath of disasters" (Sorensen et al., 2018, p. 2). This is because disasters, displacements, and income-related stress positively correlate with an increase in intimate partner violence (Enarson, 2012). When extreme weather events occur, women experiencing intimate partner violence likely have fewer options to leave violent or unhealthy relationships due to the erosion of community networks and a lack of income and resources. Climate change can exacerbate domestic violence and sexual exploitation because it removes safety nets and cuts victims off from networks that served as a safety check for women and children living in abusive or unhealthy environments. Unsafe weather conditions, infrastructure breakdown, and emergency declarations create additional challenges for victims of domestic violence to reach shelters, access a phone, or ask for help from family, friends, or community social workers. A 2009 study found that in the year following Hurricane Katrina the daily rate of gender-based violence incidents increased from 4.6 to 16.3 per 100,000 residents (Anastario et al., 2009).

Extreme weather events that disrupt living situations lead to desperation and uncertainty, particularly for single-income households. This can result in women voluntarily remaining in an abusive or unsafe environment due to limited options for relocating with family or friends during an emergency declaration (Whittenbury, 2013). Along with intimate partner violence, studies show that climate change increases the sex trafficking of women (Sheu et al., 2021). This is partly due to increased isolation from and erosion of community networks from climate events and partly due to more women entering sex work voluntarily due to limited options for work or housing after being impacted by an extreme weather event. As displacements, migration, and service disruptions increase due to climate disasters, vulnerability to human trafficking also increases. Research evidences the relationship between unstable housing and unemployment with involuntary and voluntary entry into sex work. Once a woman enters sex work, the cycle of poverty becomes more pronounced as increased dependency on others emerges and socioeconomic power declines (Gerrard, 2017). A 2024 study found that 158 million women and girls are at risk of being pushed into poverty by 2050 due to the consequences of climate change, which is 16 million more than the number of men and boys predicted to be impacted by the same events (UN Women, 2025).

Gaps in Existing Literature

The literature provides a strong overview of the inequitable distribution of climate impacts. Because the consequences of climate events are influenced by societal structures, climate change disproportionately impacts historically oppressed populations, those least responsible for carbon emissions and least equipped to adapt to a changing climate. When an emergency declaration is made, communities without resources to evacuate or relocate are more affected than communities with resources to leave disaster-affected areas. In addition, the consequences of climate change, such as extreme flooding, wildfires, failing crops, and infectious disease, are felt more strongly by communities that are already experiencing food insecurity, housing insecurity, and low socioeconomic status. Climate change not only creates new problems for historically oppressed communities but also exacerbates current vulnerabilities. As a result, low-income, single-mother households are most impacted by human-caused climate change.

In 2017, a review of the literature found that "a growing and interdisciplinary body of research has established that gender is an integral feature of social transformations associated with climate change. Gendered vulnerability to climate change is not due to characteristics intrinsic to women as a group. Rather, the social organization of women's reproductive and productive labor in different socioeconomic and cultural circumstances shapes the immediate 'factors' (e.g., poverty, education, property, and income sources) associated with women's higher risks and climate change-related burdens" (Pearse, 2017, p. 12). In 2023, *The Palgrave Handbook of Global Social Problems* published a comprehensive summary of evidence that women are more vulnerable than men to the adverse impacts of climate change, along with studies showing a positive correlation between gender equality and climate resiliency (Hailemariam et al.).

Year after year, additional studies on the gendered experience of climate change are introduced, and findings showing an unequal impact on women are published. In 2022, researcher A. O. Awiti reviewed 59 studies published from 1992 to 2022 on the gender-differentiated impacts of climate change, particularly on women in Sub-Saharan Africa, and found that "climate change affects women more negatively compared to men in five impact areas: (1) agricultural production; (2) food and nutrition security; (3) health; (4) water and energy; and (5) climate-related disaster, migration, and conflict." However, while the gap in society's collective understanding of who bears the burden of a warming planet is closing, albeit slowly, the relationship between climate change and full bodily autonomy remains an area in the literature that is not well-researched or understood in its entirety.

Over the course of four generations, the fight for gender equality has been built upon the fight for reproductive rights, namely access to contraceptives and abortion. The literature is ripe with discussion of the reproductive rights framework and studies demonstrating the positive correlation between women's socioeconomic status and abortion access (e.g., Bernstein, 2018, Foster et al., 2022; Everett & Taylor, 2024), rightfully acknowledging the criticality of full bodily autonomy in achieving full prosperity. Yet, largely absent from this discussion (and the literature) is the importance of protecting the right to have a child and the right to raise a child in a safe and healthy environment, along with the acknowledgment that full bodily autonomy cannot be achieved without the protection of both. Further, climate change as an obstacle to having and raising a healthy child is not well researched or understood.

Peppered throughout the literature are contributions that explore the relationship between environmental degradation and reproductive rights, often paraded as a discussion on reproductive justice. Addressing the link between extreme weather events and abortion access or the link between environmental justice (e.g., contaminated drinking water) and maternal health outcomes (e.g., weight of baby) is critically important research, but it is not a discussion on the relationship between climate change and reproductive justice. Many studies focus on the right to not have a child but do not acknowledge that to truly achieve reproductive justice, one must also address the right to have a child and the right to raise a child in a safe and healthy environment. Other studies confuse environmental injustice (e.g., air or water pollution) with climate change impacts (e.g., increased surface temperatures or extreme precipitation events) and focus on the adverse impacts of toxins on reproductive health outcomes. While the last couple of years have seen a few contributions on the relationship between reproductive justice and climate change (e.g., Colangelo, 2024; Sasser, 2024), the scope of the research is limited and leaves many questions surrounding the link between a changing climate and the attainment of full bodily autonomy.

According to the United Nations Population Fund, bodily autonomy is the foundation for gender equality. A 2022 study analyzed European societies over a three-hundred-year period and found a positive correlation between the protection of bodily autonomy, particularly for women, and economic prosperity. The study concluded that women's inability to realize full bodily autonomy could be a reason some European regions fell behind other regions economically, spanning as early as the sixteenth century and as recently as the nineteenth century (Baten & de Pleijt, 2022). Yet, to understand how climate change interacts with full bodily autonomy, we must first understand how the impacts of climate change act as barriers to full bodily autonomy, or, as defined by the World Economic Forum, the ability of a person (or household) to make choices about their own body without the threat of or interference by coercion, violence, or oppression.

References

Alston, M. (2013). Introducing gender and climate change: Research, policy, and action. In *Research, action, and policy: Addressing the gendered impacts of climate change* (pp. 3–14). Springer.

Anastario, M., Shehab, N., & Lawry, L. (2009). Increased gender-based violence among women internally displaced in Mississippi 2 years post–Hurricane Katrina. *Disaster Medicine and Public Health Preparedness, 3*(1), 18–26.

Arora-Jonsson, S. (2011). Virtue and vulnerability: Discourses on women, gender, and climate change. *Global Environmental Change, 21*(2), 744–751.

Asaduzzaman, M. (2015). *Livelihood vulnerability of women in the context of climate change impacts: Insights from coastal Bangladesh.* Unpublished PhD thesis, The University of Newcastle.

Awiti, A. O. (2022). Climate change and gender in Africa: A review of impact and gender-responsive solutions. *Frontiers in Climate, 4*, 895950.

Baten, J., & de Pleijt, A. M. (2022). Female autonomy generated successful long-term human capital development: Evidence from 16th to 19th century Europe. *World Development, 158*, 105999.

Bernstein, A. (2018). *Factors associated with qualifying for abortion funds: An exploratory analysis.*

Bloem, M. W., et al. (2010). Castel Gandolfo Workshop: An introduction to the impact of climate change, the economic crisis, and the increase in the food prices on malnutrition. *The Journal of Nutrition, 140*(1), 132S–135S.

Bradbear, C., & Friel, S. (2013). Integrating climate change, food prices and population health. *Food Policy, 43*, 56–66.

Colangelo, S. A. (2024). Bridging silos: Environmental and reproductive justice in the climate crisis. *California Law Review, 112*, 1255.

Dominelli, L. (2013). Gendering climate change: Implications for debates, policies, and practices. In *Research, action, and policy: Addressing the gendered impacts of climate change* (pp. 77–93). Springer.

Enarson, E. P. (2012). *Women confronting natural disaster: From vulnerability to resilience* (p. 245). Lynne Rienner Publishers.

Everett, B. G., & Taylor, C. J. (2024). Abortion and Women's Future Socioeconomic Attainment. *American Sociological Review, 89*(6), 1044–1074.

Foster, D. G., et al. (2022). Socioeconomic outcomes of women who receive and women who are denied wanted abortions in the United States. *American journal of public health, 112*(9), 1290–1296.

Gerrard, M. B. (2017). Climate change and human trafficking after the Paris agreement. *University of Miami Law Review, 72*, 345.

Henrici, J., Childers, R., & Shaw, E. (2015). *Get to the bricks: The experiences of black women from New Orleans public housing after hurricane Katrina.* Institute for Women's Policy Research.

IWPR. (2010). Women, disasters, and hurricane katrina. Institute for women's policy research. https://iwpr.org/wp-content/uploads/2020/09/D492.pdf

Klepp, S. (2017). Climate change and migration. In *Oxford research encyclopedia of climate science.*

Lane, R., & McNaught, R. (2009). Building gendered approaches to adaptation in the Pacific. *Gender & Development, 17*(1), 67–80.

MacDonald, R. (2005). How women were affected by the tsunami: A perspective from Oxfam. *PLoS Medicine, 2*(6), e178.

Neumayer, E., & Plümper, T. (2007). The gendered nature of natural disasters: The impact of catastrophic events on the gender gap in life expectancy, 1981–2002. *Annals of the Association of American Geographers, 97*(3), 551–566.

Pachauri, R. K., & Meyer, L. A. Fifth assessment report of the intergovernmental panel on climate change, ed. RK Pachauri and LA Meyer (Geneva, Switzerland: IPCC, 2014), 1–151.

Paterson, J., et al. (2014). Health care facilities resilient to climate change impacts. *International Journal of Environmental Research and Public Health, 11*(12), 13097–13116.

Pearse, R. (2017). Gender and climate change. *Wiley Interdisciplinary Reviews: Climate Change, 8*(2), e451.

Sasser, J. S. (2024). At the intersection of climate justice and reproductive justice. Wiley Interdisciplinary Reviews: Climate Change, 15(1), e860.

Sheu, J. C., et al. (2021). Potential impact of climate change on human trafficking: A narrative review. *The Journal of Nervous and Mental Disease, 209*(5), 324–329.

Sorensen, C., et al. (2018). Climate change and women's health: Impacts and policy directions. *PLoS Medicine, 15*(7), e1002603.

Tacoli, C., et al. (2015). *Urbanization, rural-urban migration, and urban poverty.* Human Settlements Group, International Institute for Environment and Development.

Tesla, B., et al. (2018). Temperature drives Zika virus transmission: Evidence from empirical and mathematical models. *Proceedings of the Royal Society B, 285*(1884), 20180795.

U. N. Women (2025). The gender snapshot 2025. https://www.unwomen.org/en/digital-library/publications/2025/09/progress-on-the-sustainable-development-goals-the-gender-snapshot-2025

Wang, J. (2010). Food security, food prices and climate change in China: A dynamic panel data analysis. *Agriculture and Agricultural Science Procedia, 1*, 321–324.

Whittenbury, K. (2013). Climate change, women's health, wellbeing, and experiences of gender-based violence in Australia. In *Research, action, and policy: Addressing the gendered impacts of climate change* (pp. 207–221). Springer.

World Health Organization. (2008). *Social determinants of health (No. SEA-HE-190).* WHO Regional Office for South-East Asia.

Zottarelli, L. K. (2008). Post-hurricane Katrina employment recovery: The interaction of race and place. *Social Science Quarterly, 89*(3), 592–607.

CHAPTER 2

Climate Change and Rising Living Costs

Barriers to bodily autonomy come in many forms. From forced birth to involuntary marriage to obligatory servitude, the realization of full bodily autonomy can be hindered by cultural practices as well as criminal enterprises. While barriers to bodily autonomy are often viewed from the vantage point of reproductive healthcare and modern slavery, included in the ability to realize full bodily autonomy is the ability to meet one's bodily needs, or the ability to access food, water, housing, and all the essentials a person needs to survive and reproduce. Accordingly, if an individual is unable to access adequate healthcare, nutritious food, or safe housing, then the individual cannot meet one's basic needs nor freely reproduce and, as a result, does not have full bodily autonomy. As discussed in Chap. 1, the consequences of climate change, measured by increased temperatures, extreme weather events, and changes in soil moisture content, can impede access to proper food, housing, and healthcare. Therefore, if a changing climate is making it harder for individuals to access food, housing, and healthcare, is climate change also increasing barriers to freely reproduce and realize full bodily autonomy?

This chapter discusses the findings of a 2024 study that explored the link between a warming planet and access to the basic necessities people need to survive and reproduce. To increase society's collective understanding of the relationship between climate change and the cost of housing, food, and healthcare, all of which are critical for maintaining the

S. Haley Knowles, *Diminished Prosperity*,
https://doi.org/10.1007/978-3-032-07068-5_2

health of families, communities, and economies, I conducted a quantitative analysis, leveraging secondary data sources, to explore the link between a changing climate, measured by changes in soil moisture content levels, increased surface temperatures, and higher occurrences of extreme weather events, and obstructions to bodily autonomy, measured by increases in food prices, housing costs, and healthcare expenditures. Using a fixed-effects linear regression model, I analyzed state-level panel data for the 48 states that comprise the contiguous United States over a 25-year period, from 1997 to 2021, to develop three models to examine the relationships between environmental, social, and economic conditions among the variables. Data from the Bureau of Economic Analysis Consumer Price Expenditure (CPE) Index by State, a consumer expenditure index that tracks consumer spending in urban areas at the state level and is organized by expenditure type, is adjusted for inflation using the United States Bureau of Labor Consumer Price Index (CPI) before being used in the modeling. Economic recession, based on the GDP-based recession indicator from Federal Reserve Economic Data (FRED), is included as a control variable in each model, represented by the number of quarters spent in economic recession each calendar year on a scale of 0 to 4.

Model 1: The Relationship Between Soil Moisture and Food Costs

My first hypothesis considers how one component of climate change, measured as the level of moisture found in soil during agricultural planting and harvesting seasons, impacts one component of bodily autonomy, defined as the cost of food. In model 1, I examine the link between climate change and food insecurity by exploring the relationship between the average annual change in soil moisture content levels, driven by a changing climate and measured by both drier conditions caused by prolonged drought and wetter conditions caused by increased flooding, and the price of consumer goods, driven by agricultural yields and commodity prices, and measured by annual food expenditures, defined as food and beverages purchased for off-premises consumption, which effectively excludes expenditures purchased at restaurants, fast food joints, cafes, and bars. Table 2.1 summarizes model 1's hypotheses and variables.

For the first part of model 1, the independent variable is the annual average planting soil moisture content level and is based on data from

Table 2.1 Model 1 hypothesis and data

Hypothesis 1	*Independent variable*	*Dependent variable*	*Controls*
H1: As soil moisture levels increase during planting and decrease during harvest, the cost of food will increase.	**IV1a:** Average monthly soil moisture level during planting season **IV1b:** Average monthly soil moisture level during harvesting season	**DV1:** Average annual food expenditures	Economic recession
Variable description	1a: Annual average planting soil moisture content at the state level during the month of April 1b: Annual average harvest soil moisture content at the state level during the month of October	Annual cost of food, or the average cost of food and beverage expenditures at the state level, defined as those purchased for off-premises consumption combined with the cost of food produced and consumed on farms, divided by the population	Number of quarters spent in economic recession at the national level for each calendar year on a scale of 0 to 4

National Oceanic and Atmospheric Administration (NOAA), which maintains a database on average soil moisture levels throughout the United States based on 344 land surface monitoring sites and estimated by a one-layer hydrological model. Data from NOAA's surface monitoring sites was analyzed by averaging the daily measurements for the month of April of all surface monitoring sites located within a state to produce an annual April soil moisture content level measurement for each state across the contiguous United States. The end result is the average soil moisture content level during the month of April measured in millimeters by soil depth and organized at the state level by year for the years 1997 to 2021. The month of April was chosen for planting season as it is the month most often associated with planting activities in agricultural communities.

Figure 2.1 shows the state-level difference for the average soil moisture content level during the month of April measured in millimeters by soil depth for the years 1997, 2010, and 2021. Overall, soil moisture content during planting season has been steadily declining over the last 25 years, with the variance across the contiguous United States becoming more pronounced. In 1997, the average soil moisture content level across the contiguous United States during the month of April was 468.5 millimeters. Twenty-five years later, the average soil moisture content level during the same month of the year dropped to an average of 420 millimeters. During planting season, Nevada has historically had the lowest levels of soil moisture, with Rhode Island and Tennessee having the highest levels. However, in 2021, the State of Mississippi had the highest soil moisture content level during planting season at 628.5 millimeters, with Nevada having the lowest at 84.3 millimeters. Sharp decreases in several states, including California, North Dakota, South Dakota, and Idaho, took place over the last few years with decreases as high as 155 millimeters over just a 12-month period.

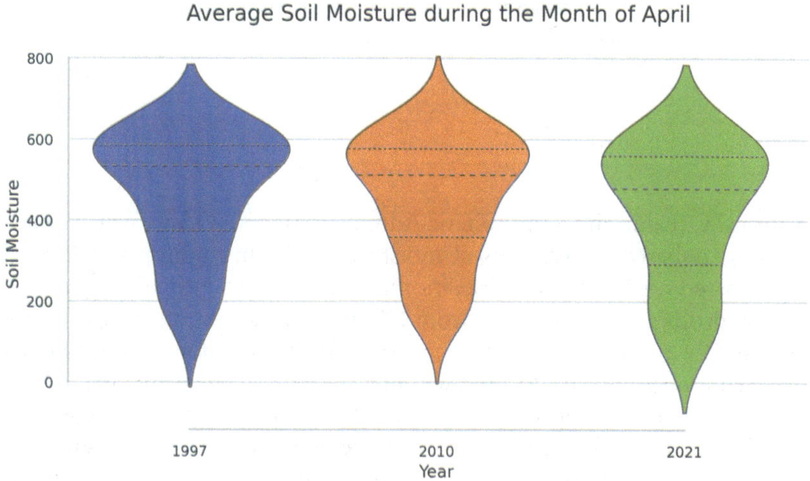

Fig. 2.1 Average soil moisture during the month of April

For the second part of model 1, the independent variable is the average harvesting soil moisture content level and is calculated using the same approach as model 1a but for the month of October. The month of October was chosen for the harvest season as it is the month most often associated with harvesting activities in agricultural communities. Figure 2.2 shows the state-level difference for the average soil moisture content level during the month of October, measured in millimeters by soil depth for the years 1997, 2010, and 2021. Based on data for soil moisture content for the month of October, soil moisture is steadily increasing year over year during the harvest season, which may correlate with the wet season that often brings with it extreme precipitation events. For example, in 1997, the average soil moisture content level across the contiguous United States during harvest season was 362.1 millimeters, and by 2021, levels during the month of October increased to an average of 377.7 millimeters. In 2021, the State of Massachusetts had the highest soil moisture content level during harvest season at 581.3 millimeters, and the State of Nevada had the lowest at 66.2 millimeters.

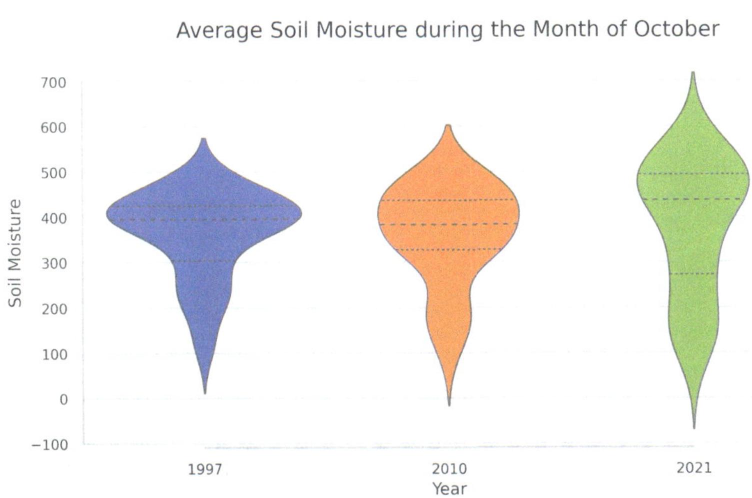

Fig. 2.2 Average soil moisture during the month of October

Model 1's dependent variable is the annual cost of food and is calculated using the BEA CPE Index by State for all food and beverage expenditures purchased for "off-premises" consumption, along with food produced and consumed on farms. The annual dollar amount in millions of current dollars at the state level is organized by year for the years 1997 to 2021, adjusted for inflation using the CPI, and divided by the corresponding state population based on data from FRED, to generate an average cost of food. Excluded from the dependent variable is any money spent by consumers on food for consumption outside of the home, such as at restaurants, bars, diners, fairs, airports, or any location where food or beverage was purchased to be consumed at the location where it was purchased.

Figure 2.3 shows the state-level difference for annual food expenditures purchased or produced to consume at home, measured in millions of dollars, for the years 1997 through 2021. Overall, the average cost of food increased significantly over the 25-year period, from an average rate of $4.2 million per 1000 people in 1997 to $17 million per 1000 people in 2021. The States of Maine, Vermont, New Hampshire, Massachusetts,

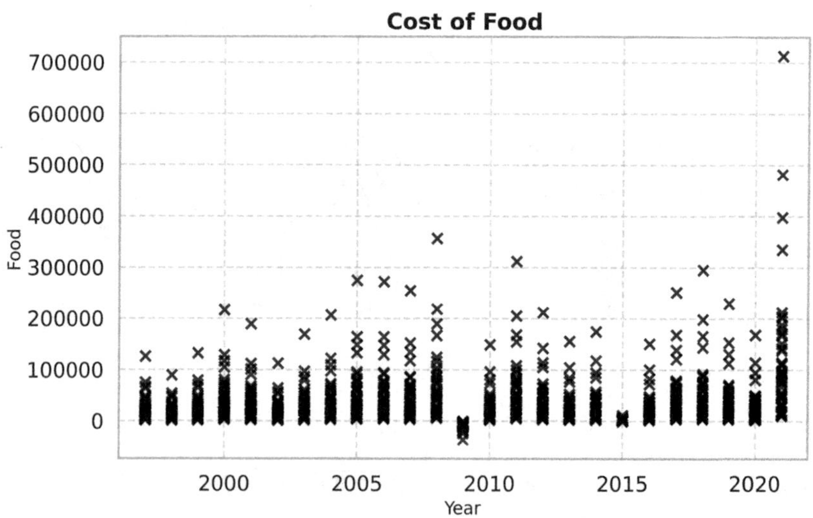

Fig. 2.3 Annual cost of food

and Colorado had the highest rates of food costs over the 25-year period, with the highest costs across all states occurring in 2021, followed by 2008 and 2011, which all correlate with a national economic recession. However, a drastic decrease in the cost of food occurred across all states during the years 2001, 2009, 2015, and 2020, with 2009 showing the lowest rate of food costs across the contiguous United States.

Figures 2.4, 2.5, and 2.6 show the state-level difference for the annual rate of food expenditures purchased or produced to consume at home for the years 1997 versus 2021. The darker the color of red, the higher the dollar amount. The lightest color of red indicates the lowest cost of food. Overall, the cost of food has dramatically increased over the last 25 years in every state with variance remaining fairly steady. Notably, the rate of food costs increased the most in New York (381.17%), Maine (369.78%), California (365.89%), Colorado (348.60%), and Rhode Island (348.20%). In 2021, the top ten states with the highest cost of food were Maine, Vermont, New Hampshire, Massachusetts, Colorado, Oregon, Wyoming, Connecticut, Montana, and Washington, with an average annual cost of $19,564 per capita. In 2021, the States of Oklahoma, Tennessee, Utah, and Arkansas had the lowest cost of food, with an average annual cost of $14,109 per capita.

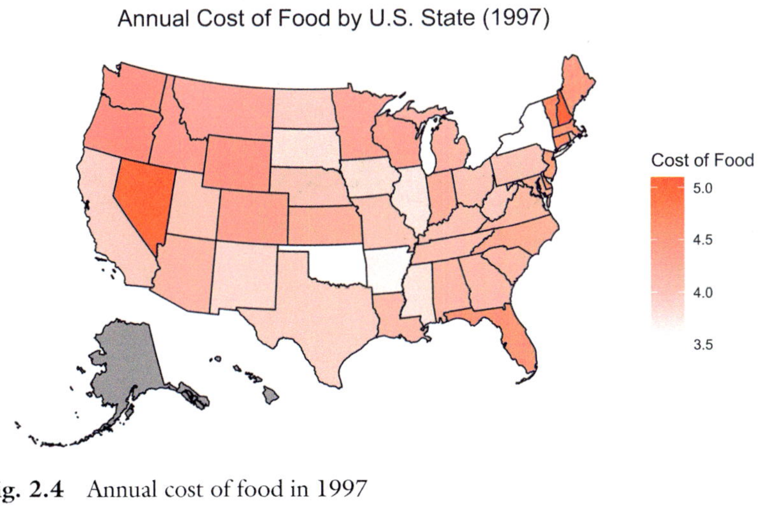

Fig. 2.4 Annual cost of food in 1997

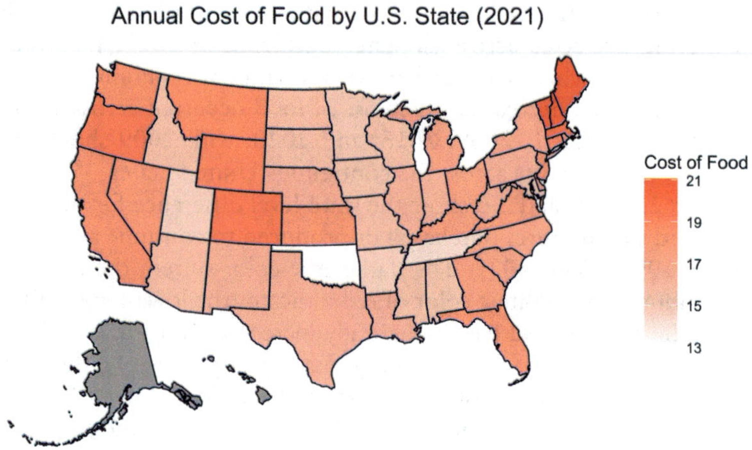

Fig. 2.5 Annual cost of food in 2021

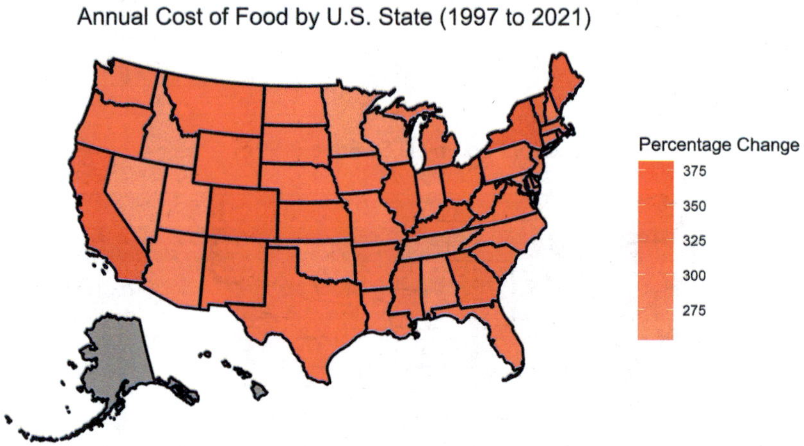

Fig. 2.6 Change in annual cost of food

The first part of model 1 presents the effects of average soil moisture content levels on annual food expenditures, controlling for economic recession, and finds that the average soil moisture content level during the month of April, known most commonly as planting season, is significant. For every millimeter the soil moisture content level decreases during the month of April, the cost of food increases by approximately \$5 per capita. However, the second part of model 1, or the relationship between the level of soil moisture content during harvesting season and food expenditures, is found to be only marginally significant. So, while for every millimeter the soil moisture content level increases during the month of October, often associated with harvesting and the end of the dry season, the cost of food is expected to increase by approximately \$4 per capita, the extent of and reasons behind the relationship between soil moisture content during the month of October and food expenditures are less clear based on the modeling, data, and variables used in this study. Figure 2.7 illustrates the relationship between soil moisture (during both planting season in the Spring and harvesting season in the Fall) and food expenditures at the state level across years 1997–2021.

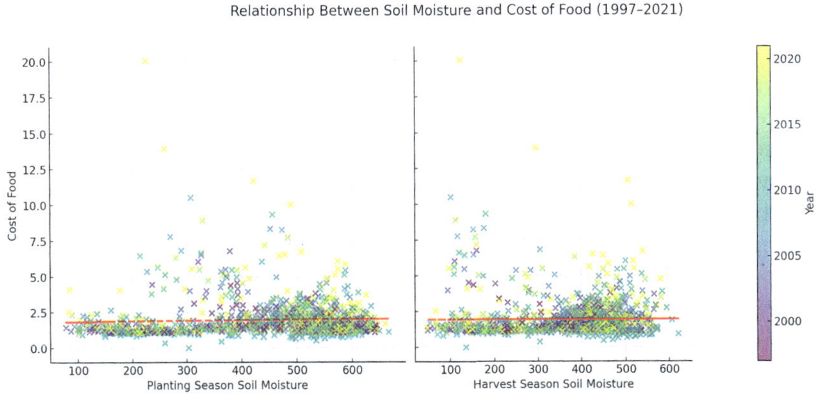

Fig. 2.7 Relationship between soil moisture and food costs

While an increase of $4 or $5 annually per person may seem trivial, overall, food is getting more expensive. The USDA reports that the cost of groceries has increased around 1.2% each year over the last 30 years but went up nearly 20% in 2022 compared to 2021. In addition, the same agency found the cost of groceries was about 5% higher in 2023 than in 2022, with the 20-year historical level of retail food price inflation around only 2.5%. From 2019 to 2023, the amount of money consumers in the United States spent on food rose by 25% compared to a 19.2% increase for all other items tracked by the BEA CPE. Further, while the average decrease in soil moisture during planting season is less than 50 millimeters from 1997 to 2021, not all soil across the contiguous United States is created equal. For example, several states where staple crops are grown saw steep decreases in soil moisture over just a one-year period, such as California, which decreased 84 millimeters from 2020 to 2021, equating to an annual increase of $2100 in the cost of food for a family of five. In addition, during the same year, North Dakota decreased 122 millimeters, and South Dakota decreased about 100 millimeters, which adds approximately $3050 and $2500 respectively to the annual cost of groceries for a family of five living in one of these states due to soil moisture content decreases during the month of April alone.

The results of this study highlight what every farmer already knows—when soil is too dry to plant due to the gradual decrease in soil moisture year over year, or too wet to harvest due to extreme weather events that often accompany harvest in the Fall, agricultural yields are likely to decrease, which triggers a chain reaction from high commodity prices to higher prices at the grocery store. Further, the adverse impacts to healthy soil from a changing climate do not appear to be subsiding, but may actually be getting worse. NASA reports that between 2000 and 2020, global soil moisture content levels, measured throughout all 12 months of each year, decreased at a rate of 1284 kilograms per millimeter annually and are expected to continue on the same trajectory. Based on these findings, if soil moisture levels are expected to decrease globally by 1284 kilograms per millimeter each year, and if for each millimeter that soil moisture decreases during planting season across the United States, consumers are expected to spend an additional $5 per capita on the cost of groceries, then consumers could see significant increases in food expenditures each year from soil moisture decreases alone.

Model 2: The Relationship Between Disaster Declarations and Housing Costs

My second hypothesis considers how a second component of climate change, measured by extreme weather events, impacts a second component of bodily autonomy, defined as the cost of housing. In model 2, I explore the link between extreme climate events and increased housing costs. As evidenced by the literature, climate disasters are linked to increased housing costs through a decrease in housing supply following an extreme weather event, caused by the resulting property damage, and an increase in housing demand, caused by the resulting displacement from the loss of homes and the erosion of social safety nets. Table 2.2 summarizes model 2's hypothesis and variables.

To calculate model 2's independent variable, data is pulled from the Federal Emergency Management Agency (FEMA) Disaster Declarations index, which, based on the Robert T. Stafford Disaster Relief and

Table 2.2 Model 2 hypothesis and data

Hypothesis 2	*Independent variable*	*Dependent variable*	*Controls*
H2: As occurrence of extreme weather events increase, the cost of housing will increase.	**IV2:** Annual disaster declarations approved for federal assistance	**DV2:** Average annual housing and utilities expenditures	Economic recession
Variable description	Annual number of FEMA disaster declarations at the state level approved for public and/or individual assistance programs	Annual cost of housing, or the average cost of housing and utilities expenditures at the state level, defined as those monetary rents paid by tenants for tenant-occupied housing, the rental value of farm dwellings, and spending on group housing, combined with the cost of utilities, consisting of water, electricity, and gas, divided by the population	Number of quarters spent in economic recession at the national level for each calendar year on a scale of 0 to 4

Emergency Assistance Act, grants the President of the United States authority to approve federal assistance for United States residents impacted by a disaster. However, the independent variable for model 2 only includes disaster declarations approved for Public Assistance, which provides federal funding to state and local governments and nonprofit organizations for the cost of disaster-related debris removal, emergency protective measures, and permanent repair work to damaged or destroyed infrastructure; or Individual Assistance, which provides financial and direct services to eligible individuals and households affected by a disaster. The data is organized at the state level by year for the years 1997 to 2021.

Figure 2.8 shows the state-level difference for annual FEMA disaster declarations approved for federal assistance for the years 1997 through 2021, totaling 3170 FEMA disasters. In 1997, the total number of disaster declarations across the contiguous United States was 73. In 2021, the number jumped to 217 declarations. However, only a handful of states experienced 20 or more disasters per year, with over half reaching that number within just the past decade. Outliers include Texas, California, Oregon, Oklahoma, Washington, Florida, and Colorado. The State of Texas had the most declarations with 58 disasters in 2011. California reached 57 disasters in 2018 and 2020. Oregon experienced 39 disasters in 2020. The median at the state level over the last 25 years is two disaster declarations each year.

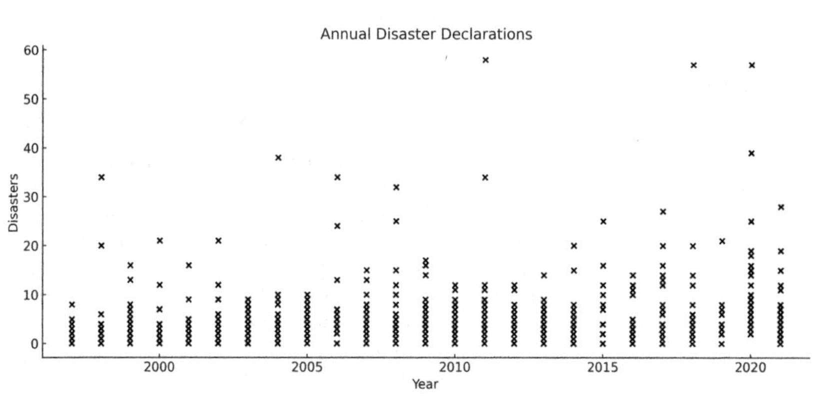

Fig. 2.8 Annual disaster declarations

Model 2's dependent variable is the average amount of money spent on housing and utilities each year at the state level and is calculated using the BEA CPE Index by State. Housing expenditures include the total amount of monetary rents paid by tenants for tenant-occupied housing, the rental value for owner-occupied dwellings measured as the income the home-owner could have received if the house had been rented to a tenant, the rental value of farm dwellings, and total spending on group housing. Household utilities consist of water, sanitation, electricity, and gas. The annual dollar amount in millions of current dollars at the state level is organized by year for the years 1997 to 2021, adjusted for inflation using CPI, and divided by the corresponding state population based on data from FRED to generate an average cost of housing.

Figure 2.9 shows the state-level difference for annual housing and utilities expenditures measured in millions of dollars, for the years 1997 through 2021. Outliers include Colorado, California, Oregon, Washington, Massachusetts, Nevada, and Arizona. The year 2021 saw the highest housing costs, followed by 2008, with the cost of housing being highest over the 25-year period in the States of Colorado, Massachusetts, California, Washington, New Jersey, and Florida, with an average annual cost of $46,915 per capita, or a monthly housing payment of $3910. Ironically, housing costs plummeted to the lowest levels over the same period in 2009 in some of the same states that also showed the highest rates of housing costs, namely New Jersey, Florida, and Massachusetts; showing how quickly and sharply the housing market can fluctuate based on impacts to supply and demand.

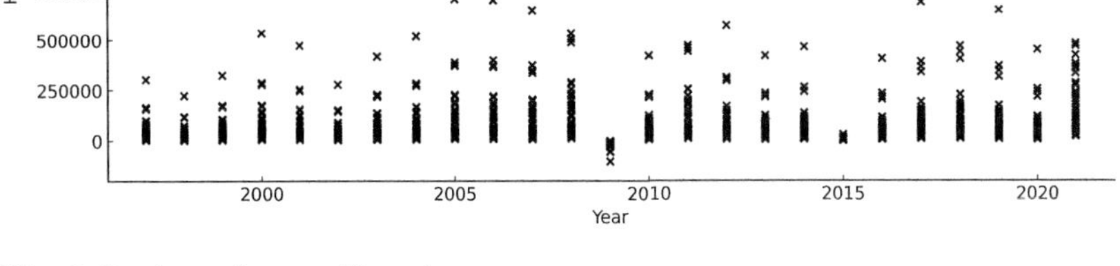

Fig. 2.9 Annual cost of housing

Figures 2.10, 2.11, and 2.12 show the state-level difference in the annual rate of housing expenditures for the year 1997 versus 2021. The darker the color of red, the higher the dollar amount. The lightest color of red indicates the lowest cost of housing. Similar to food costs, the cost of housing has dramatically increased over the last 25 years in every state with

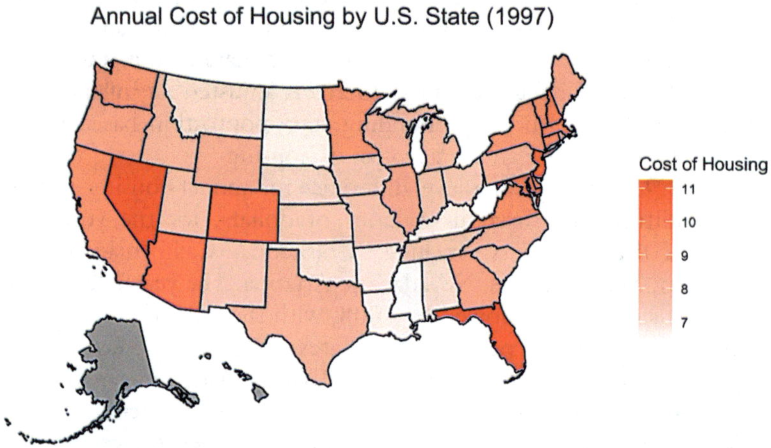

Fig. 2.10 Annual cost of housing in 1997

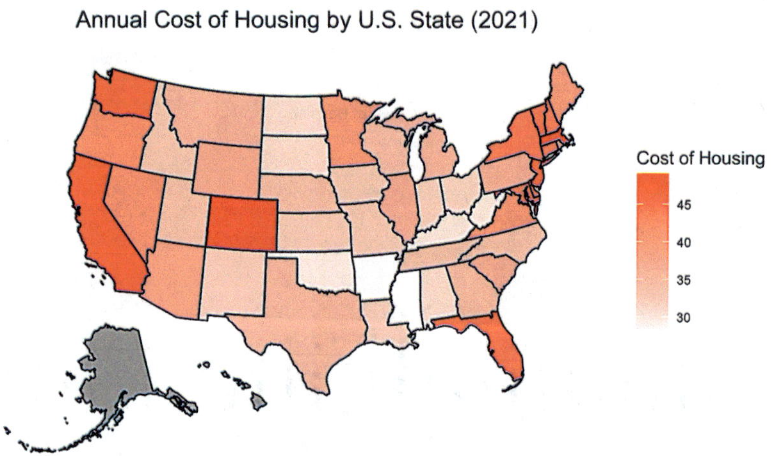

Fig. 2.11 Annual cost of housing in 2021

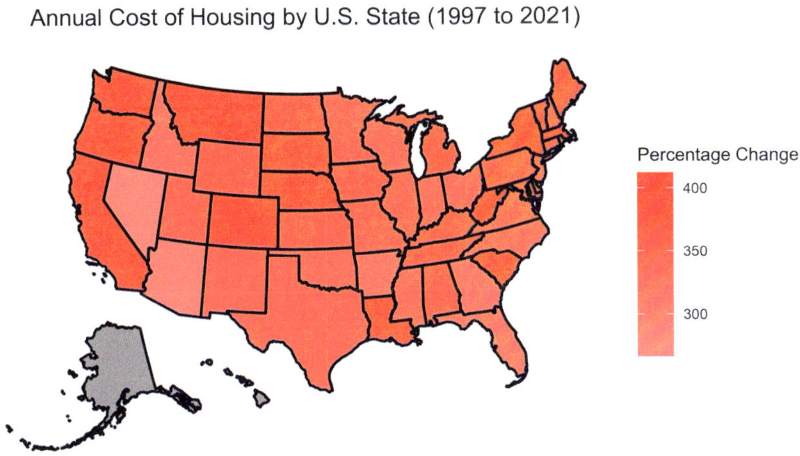

Fig. 2.12 Change in annual cost of housing

Washington (411.90%), Montana (410.43%), California (398.32%), Massachusetts (389.70%), and Louisiana (386.31%) showing the biggest increases. In addition, variance across housing costs at the state level significantly increased from 1.62 in 1997 to 34 in 2021.

In model 2, the number of federal disaster declarations positively correlates with the amount of money consumers spend on housing and utilities each year. Specifically, for each additional disaster declaration approved for public or individual assistance by FEMA, the cost of housing and utilities will increase by $165 per capita per year. While $165 spread across 12 months may not seem like a significant increase in housing and utilities costs, a household of five could see an increase of $825 a year for each additional disaster declaration, or an increase of $68.75 per month. Further, the State of California experienced 57 disasters in 2018 and 2020, which is an increase of 30 from the 27 disasters experienced in 2017 and an increase of 36 from the 21 disasters experienced in 2019. Based on these findings, a family of five living in California could face an increase in their annual housing costs of $24,750 in 2018 and $29,700 in 2019. Figure 2.13 illustrates the relationship between disaster declarations and the annual rate of housing expenditures at the state level across years 1997–2021.

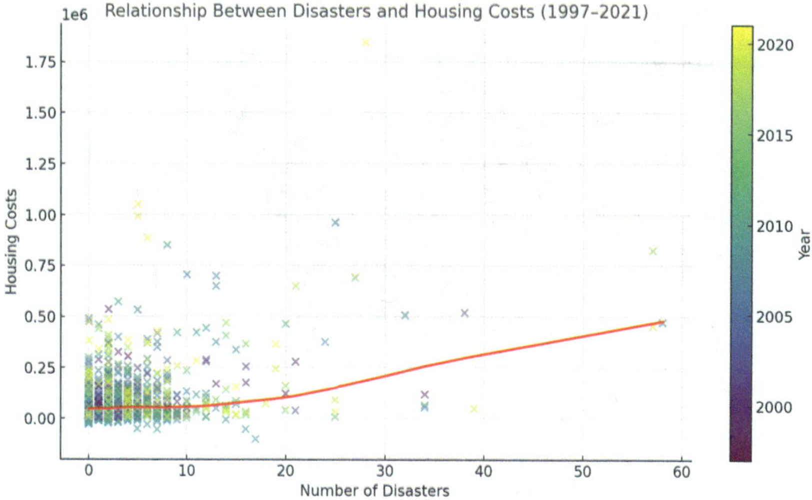

Fig. 2.13 Relationship between disasters and housing costs

Extreme climate events have been trending upwards over the past few decades, with the likelihood that disaster declarations will continue to increase year over year. According to FEMA, each year since 2019, consumers living in the United States have been experiencing more than 200 disasters, with 20 of those disasters costing more than a billion dollars each, which is a sharp increase from the average of three such disasters seen during the 1980s. Further, with 2023 seeing a record 23 climate disasters costing more than a billion dollars each, the climate-fueled destruction of homes and the resulting decrease in housing supply and increase in housing demand are expected to only get worse.

Model 3: The Relationship Between Surface Temperatures and Healthcare Costs

My third hypothesis considers how a third component of climate change, measured by surface temperatures, impacts a third component of bodily autonomy, defined as the cost of healthcare. In the third model, I examine the relationship between hotter surface temperatures and increased healthcare expenditures, defined as consumer spending on outpatient services,

Table 2.3 Model 3 hypothesis data

Hypothesis 3	*Independent variable*	*Dependent variable*	*Controls*
H3: As surface temperatures increase, the cost of healthcare will increase.	**IV3:** Average monthly temperature during hottest month of the year	**DV3:** Average annual healthcare expenditures	Economic recession
Variable description	Annual average temperatures at the state level during the month of July	Annual cost of healthcare, or the average cost of healthcare expenditures at the state level, defined as outpatient services, consisting of physician, dental, and paramedical, hospital services, and nursing home services, divided by the population	Number of quarters spent in economic recession at the national level for each calendar year on a scale of 0 to 4

including physician, dental, and paramedic services, and inpatient services, including hospital and nursing home services. Based on the literature, as temperatures increase, not only are humans more likely to suffer from heat stress, heat stroke, and other heat-related illnesses, but vectors, such as mosquitoes and ticks, are able to live longer and travel to more regions populated by humans, which increases the likelihood of vector-borne diseases being transmitted to humans. Table 2.3 summarizes model three's hypothesis and variables.

In the third model, the independent variable is the average monthly temperature during the month of July, most often associated with the hottest month of the year for the contiguous United States, and is based on data from NOAA's Climate Normals datasets, which compare typical climate conditions against a uniform 30-year period. The variable is calculated using NOAA's 344 surface monitoring sites and by averaging the daily surface temperature measurements for the month of July from all surface monitoring sites located within a state to produce an annual July surface temperature measurement for each state across the contiguous United States. The end result is the average surface temperature during the month of July organized at the state level by year for the years 1997 to 2021.

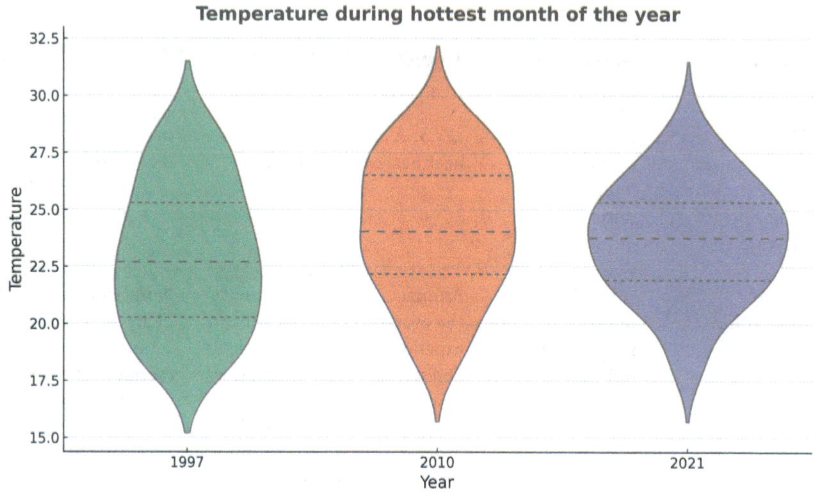

Fig. 2.14 Average surface temperature during the month of July

Figure 2.14 shows the state-level difference for the average surface temperature during the month of July, measured in degrees Celsius, for the years 1997, 2010, and 2021. Overall, temperatures are steadily increasing. In 1997, the average temperature in July was 22.9 degrees Celsius. By 2021, the average temperature rose to 23.7 degrees Celsius, an average increase of 0.08 degrees Celsius across the states, with variance decreasing from 18.32 in 1997 to 6.4 in 2021. The highest July temperature during 15 of the last 25 years occurred in three states, namely Oklahoma, Arizona, and Texas, while the lowest temperature occurred in seven states—New Hampshire, Washington, Maine, Michigan, Oregon, Vermont, and Wisconsin.

Model 3's dependent variable is based on data from the BEA CPE Index by State on annual healthcare expenditures, which are calculated based on the amount of money consumers spent on healthcare needs, including insurance premiums, deductibles, copayments, coinsurance, and out-of-pocket expenses. The dependent variable is comprised of healthcare costs in three categories: (1) outpatient services, consisting of physician, dental, and paramedical; (2) hospital services; and (3) nursing home services. The annual dollar amount in millions of current dollars at the state level is organized by year for the years 1997 to 2021, adjusted for inflation using CPI, and divided by the corresponding state population based on data from FRED to generate an average cost of healthcare.

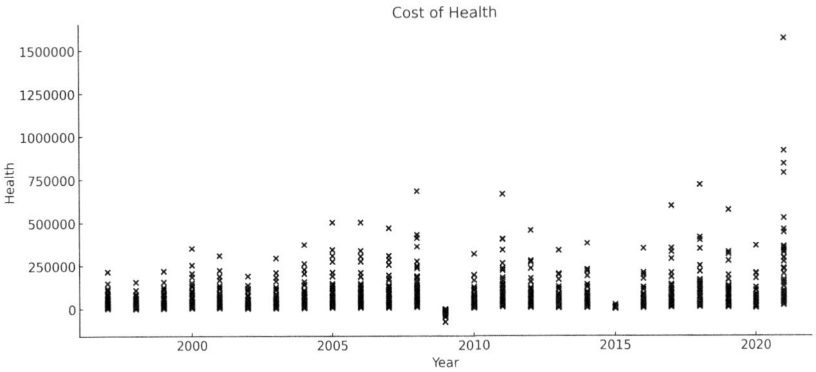

Fig. 2.15 Annual cost of healthcare

Figure 2.15 shows the state-level difference for annual healthcare expenditures, measured in millions of dollars, for the years 1997 through 2021. Outliers include the States of South Dakota, North Dakota, and Massachusetts, where the annual cost of healthcare in 2021 was \$53,962, \$46,570, and \$46,370 per capita, respectively. In contrast, for the same period, the cost of healthcare in the State of Mississippi was \$27,345 per capita. Interestingly, in 1997, the annual cost of healthcare in South Dakota was \$6,601 per capita, and \$6,024 per capita in Mississippi.

Figures 2.16, 2.17, and 2.18 show the state-level difference in the annual rate of healthcare expenditures for the year 1997 versus 2021. The darker the color of red, the higher the cost of healthcare. The lightest color of red indicates the lowest cost of healthcare. Similar to food and housing, the overall cost of healthcare has dramatically increased over the last 25 years with South Dakota (717.48%), Montana (580.84%), North Dakota (573.46%), Wyoming (557.19%), and West Virginia (549.07%) showing the biggest increases. The year 2021 saw the largest increase with the annual cost of healthcare being highest in the State of South Dakota (\$53,960 per capita) and lowest in the State of Utah (\$26,940 per capita). Over the 25-year period, variance across the states sharply increased from 0.77 in 1997 to 26.94 in 2021.

In the third model, the level of surface temperatures during the month of July, often associated with the hottest month in the United States, positively correlates with the amount of money consumers spend on healthcare during the same year. Specifically, for each degree Celsius that the

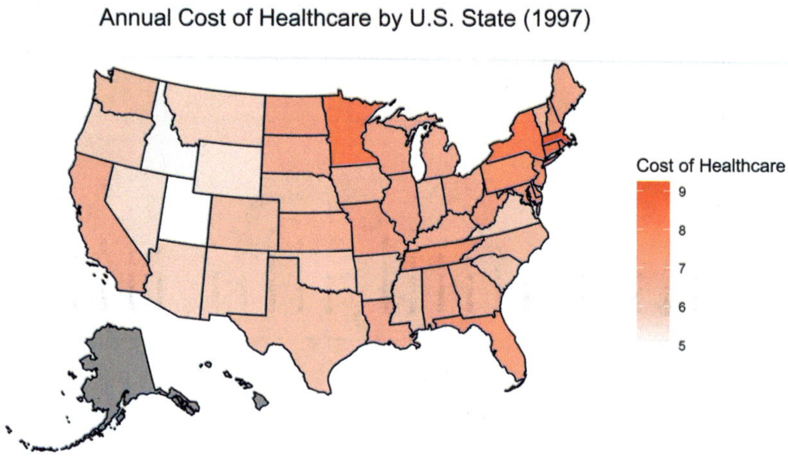

Fig. 2.16 Annual cost of healthcare in 1997

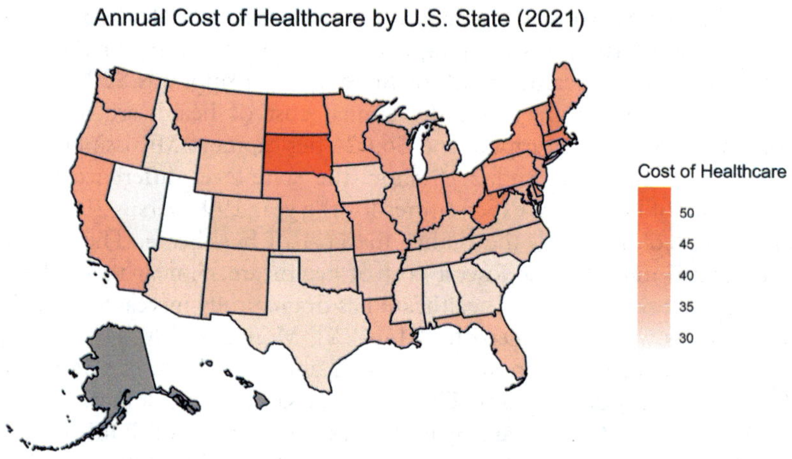

Fig. 2.17 Annual cost of healthcare in 2021

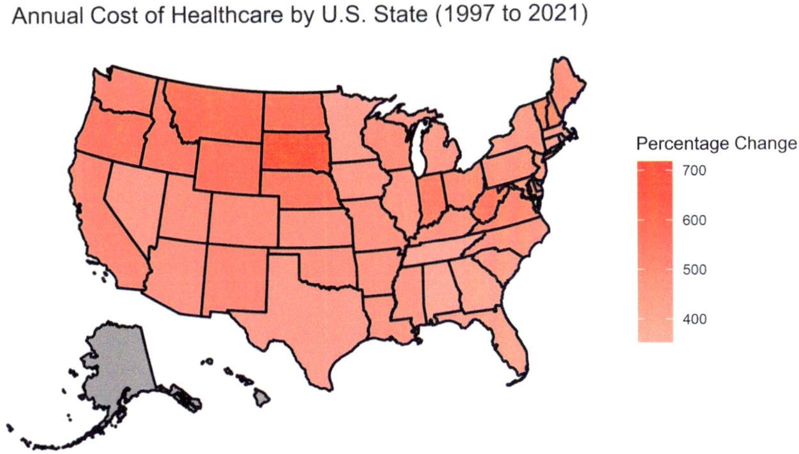

Fig. 2.18 Change in annual cost of healthcare

average temperature during the month of July increases, the cost of healthcare increases by approximately $1110 per person per year. Importantly, according to NASA, 2023 was the hottest year on record, with temperatures 1.4 degrees Celsius hotter than in 1880, when humans first started measuring and tracking surface temperatures. In addition, NASA reports that global surface temperatures have increased approximately 0.08 degrees Celsius each decade since 1880 and are expected to continue on this pathway. Figure 2.19 illustrates the relationship between July surface temperatures and healthcare expenditures at the state level across years 1997–2021.

As previously discussed, increased summer temperatures are associated with heat-related illnesses, and the month of July tends to have the highest temperatures of the year throughout the contiguous United States, along with children being out of school and spending more time outside, which makes them more susceptible to the adverse effects of high temperatures. As a result, a family of five could see an increase each decade in annual healthcare expenditures of approximately $444 per year, resulting in close to $100 more each month in healthcare costs by the end of the next 25-year period based on increased surface temperatures alone. An increase

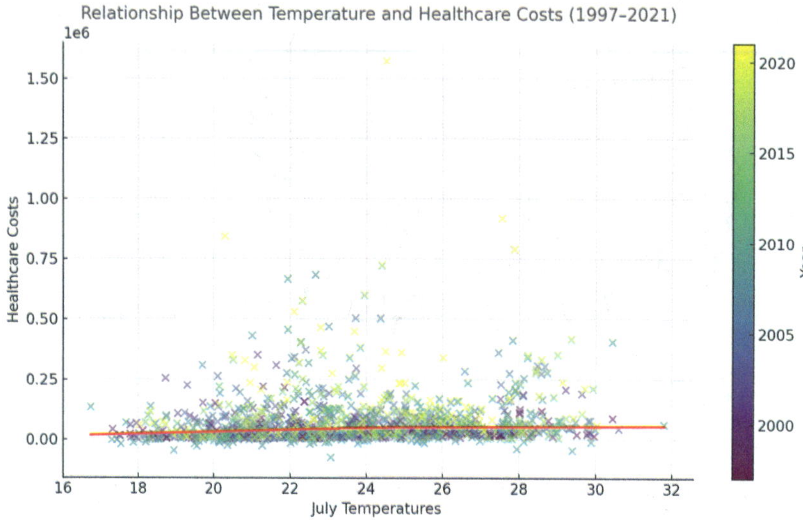

Fig. 2.19 Relationship between temperatures and healthcare costs

of this amount in healthcare expenditures, when also considering other cost-of-living increases, such as food and housing costs, could be crippling to middle- and lower-income household budgets.

This study concludes that the impacts of climate change, measured by soil moisture, surface temperature, and extreme weather events, positively correlate with increased costs of essential goods, measured by the consumer price of food, housing, and healthcare. Importantly, however, the relationship between the independent and dependent variables in each model is one of correlation, not causation. While the findings show a very significant relationship between surface temperatures and healthcare costs, extreme weather events and housing expenditures, and soil moisture and food prices, the study only evidences a correlation between the variables, not a cause-and-effect dynamic. Potential future extensions of the research include exploring other factors that could be contributing to an increase in living costs, such as Medicaid expansion, supply chain disruption, or urban industrial booms.

As climate change progresses and global temperatures continue to rise, critical soil moisture that helps grow bountiful harvests will continue to fluctuate beyond historical levels, decreasing agricultural yields and driving up the cost of food. The number of summer days with temperatures that surpass the human body's healthy threshold for heat will soar, and heat-related illnesses will gradually rise, bringing with them the rising cost of healthcare. Extreme weather events will become more frequent, wreaking havoc on households across the country while slowly deteriorating affordable housing stock. Therefore, if climate change disproportionately and adversely impacts women (as discussed in Chap. 1) while also contributing to the rising costs of food, housing, and healthcare, which are all essential to be able to have a healthy child and parent children in safe and healthy environments, does a warming planet impede the realization of full bodily autonomy and the fulfillment of fertility desires?

CHAPTER 3

Reproductive Justice and Bodily Autonomy

The disproportionate impacts of climate change on women and other historically oppressed populations infringe upon the principles of reproductive freedom and full bodily autonomy. While various theories help to evaluate freedom, choice, and agency, the reproductive justice framework provides the best lens to view the impact of a changing climate on women and the fulfillment of fertility desires. The framework provides a set of principles to examine the unique impacts of climate change on women: the right to bodily autonomy, the right to not have a child, the right to have a child, and the right to parent in safe and healthy environments (Ross & Solinger, 2017). Any barrier infringing upon these rights is an attack on freedom and considered a reproductive injustice. These principles are discussed through a historical recalling of American slavery and the origins of gynecology, eugenics, and reproductive coercion, case law, the reproductive justice movement, the theory of intersectionality, relational autonomy, and continued bodily autonomy regulation.

Reproductive Coercion and Eugenics

A long history of reproductive injustice is linked to American slavery. Because enslaved men and women were considered property of their slave owners, they were stripped of all autonomy, including control over reproductive choices and the well-being of their children. Enslaved women

S. Haley Knowles, *Diminished Prosperity*,
https://doi.org/10.1007/978-3-032-07068-5_3

were often forced to give birth to multiple children, usually more than they would have had of their own accord, as a method of replenishing the slave owner's property (Roberts, 1999). As a result, many enslaved women suffered from complications due to obstructed childbirth. One such complication was a vesicovaginal fistula, a common nineteenth-century complication of childbirth in which a tear between the uterus and bladder causes urine leakage and infertility. At the time, there was no cure for vesicovaginal fistula, which caused slave owners to become increasingly concerned over the loss of an enslaved woman's ability to provide them with new property (Pumphrey, 2014).

J. Marion Sims, an affluent white man and doctor, began experimenting on enslaved Black women to pioneer a surgical method to repair vesicovaginal fistulas. From 1845 to 1849, Sims perpetuated many of the first incidents of reproductive injustice documented in the United States by performing dozens of experimental surgeries on at least 12 Black women without the use of anesthesia. While anesthesia was available at the time, Sims did not provide anesthesia to the women due to his racist belief that Black people do not feel pain. Sims eventually mastered the surgery, cured vesicovaginal fistulas, and became known as the father of gynecology (Pumphrey, 2014). It is here that the relationship between consumerism, property, and reproductive injustice is first introduced.

The United States has a long, violent history of reproductive coercion, forced birth, forced sterilization, and reproductive experimentation on historically oppressed communities. Following the abolishment of slavery and the end of the Civil War, eugenics began to gain popularity as a way to control the reproduction of populations deemed less desirable by society. Beginning in the late 1800s, the eugenics movement provided the foundation for over a century of forced sterilizations of those deemed to possess undesirable physical or mental characteristics (Roberts, 1999). When contraceptive pills were introduced in the 1950s, a public housing project in Puerto Rico was selected for the first large-scale human trial. Between 1929 and 1974, North Carolina's Eugenics Board sterilized close to 7,600 people, the majority of whom were low-income, Black women. From 1970 to 1976, up to 50% of Indigenous women living in the United States were sterilized, often without their consent. These incidents of reproductive injustice were fueled by racism and consumerism, as well as public policy (Ross & Solinger, 2017).

In 1927, the Supreme Court of the United States ruled in *Buck v. Bell* that compulsory sterilizations are constitutional. The case stemmed from the state-mandated sterilization of a young woman named Carrie Bell by the State of Virginia. Bell was institutionalized after the state determined her to be "feeble-minded," which resulted in her sterilization. Shockingly, while explaining his support for the Virginia state law and in response to Bell's family history of low income and lack of formal education, Supreme Court Justice Oliver Wendell Holmes declared that "three generations of imbeciles is enough." Following the historic ruling, the next five decades saw the mass sterilization of up to 70,000 people in the United States, possibly more, mostly low-income communities and people of color. Ironically, these incidents of reproductive injustice collided with the start of the reproductive rights movement and, in many ways, were exacerbated by pro-birth control advocates (Roberts, 1999).

The intersection of eugenics and the reproductive rights movement began in the 1920s. Margaret Sanger, a pro-birth control advocate, eugenicist, and founder of Planned Parenthood, advocated for racial betterment through the increased use of birth control methods. Following the success of her birth control clinic in Harlem, Sanger founded the Negro Project, which operated from 1939 to 1942. The project paraded itself as a provider of education for communities with limited resources, while targeting low-income Black communities located in the rural South for the distribution of birth control. Critics of the Negro Project highlight Sanger's methods as a form of reproductive coercion to discourage low-income Black communities from reproducing. Ironically, the same birth control celebrated for liberating affluent white women from unchosen reproductive destinies was weaponized to control the reproductive choices of low-income communities and women of color (Roberts, 1999).

At the end of 1990, the Food and Drug Administration approved Norplant, the first subdermal implantable contraceptive device, to prevent pregnancies for up to five years. As a result, Norplant was used by policymakers, judges, social workers, and nonprofit organizations to coercively sterilize low-income communities and communities of color by mandating the use of Norplant for welfare recipients (Roberts, 1999). Between the years 1991 and 1992, 14 states passed laws tying Norplant to welfare benefits, all with the goal of reducing the number of children born to parents receiving Welfare benefits. In addition, since the device could only be implanted and removed by a doctor, lawmakers further incentivized Medicaid recipients by authorizing Medicaid to cover the removal of the device only at the end of the five-year period. Years later, the harmful side effects of the birth control device were discovered, and Norplant was taken off the market, but not before causing irreparable harm to many low-income and minority individuals, including infertility (Licata, 2020).

While history shows the irreparable harms of weaponizing birth control, protecting safe and affordable access to contraceptives is vital for reproductive freedom. Further, although *Buck v. Bell* has not formally been overturned, reproductive freedom has celebrated several victories over the past seven decades. The 1965 Supreme Court decision in *Griswold v. Connecticut* struck down a state prohibition against the prescription, sale, or use of contraceptives and held that the Constitution guarantees a "right to privacy" for married individuals in decisions around family planning. In 1972, unmarried individuals gained the right to birth control in *Eisenstadt v. Baird.* In 1973, the landmark abortion rights case, *Roe v. Wade*, was decided, legalizing all abortions up to the point that a fetus is determined viable. However, while these victories for the reproductive rights movement took center stage, incidents of reproductive oppression continued, particularly for those in low-income households and without access to funding for healthcare.

In the 1980 case of *Harris v. McRae*, the Supreme Court rejected a challenge to the Hyde Amendment, which, since being adopted by Congress in 1976 and amended in 1993 to add additional exceptions, bans the use of federal Medicaid funds to cover abortion except in incidents when the mother's life is in danger or in cases of rape or incest. In the 1991 case of *Rust v. Sullivan*, the Court also upheld the gag rule, which was adopted in 1984 to prohibit abortion counseling and referral by family planning programs funded under Title X of the federal Public Health Service Act. In 1992, reproductive freedom was dealt another blow in *Planned Parenthood of Southeastern Pennsylvania v. Casey.* Although the Court preserved constitutional protection for the right to choose, it adopted a weaker test for evaluating restrictive abortion laws. Under the new undue burden test, state regulations on abortion are allowed if they do not place a substantial obstacle in the path of a patient seeking a legal abortion. However, what is considered substantial is left for interpretation. This has resulted in increased barriers to care, impacting mostly low-income communities and women of color (Roberts, 1999).

On June 24, 2022, the Supreme Court ruled that the U.S. Constitution does not confer a right to abortion in *Dobbs v. Jackson*, overturning *Roe v. Wade* and returning power to legislate abortion access to the states. Since then, dozens of cases have been litigated in state and federal courts over a citizen's right to access abortion care, with dozens more pending. While access to safe, legal abortion is vital for healthy communities and any abortion restriction impedes on full bodily autonomy, the modern reproductive rights movement continues to target abortion access as its primary issue. Accordingly, the scope of historical reproductive freedom cases is limited mostly to abortion access and lacks a holistic approach to liberation for those who are most oppressed in their reproductive choices, particularly low-income women and women of color.

The Birth of the Reproductive Justice Movement

To fully understand what constitutes reproductive injustice, one must first understand how to recognize reproductive justice. In June of 1989, while white feminists were busy leading the reproductive rights movement, centering white women's issues and exacerbating the oppression of marginalized women, over 1,500 Black women gathered at Spellman College in Atlanta, Georgia. It was the first national conference on Black women's health issues in response to centuries of reproductive oppression experienced by Black women and the communities they are a part of and represent. While the rallying call to remediate women's reproductive oppression was not new, the women's movement historically focused on white women's issues and forced Black, Indigenous, Hispanic, and queer women's issues to the side (Ross et al., 2016).

White women had taken up the torch against patriarchy while leveraging the power of white supremacy, capitalism, and purity culture to distinguish affluent, well-educated, cisgender, heterosexual white women from others. Because the women's movement is largely controlled by affluent white women, the injustices of population control and forced sterilization were not an agenda item. For centuries, Black women suffered the most from reproductive oppression and yet were largely excluded from organizing around reproductive rights. The irony of this was not lost on those who gathered in Atlanta during the Summer of 1989. The conference led to the creation of the National Black Women's Health Project, which laid the foundation for what would become an organizing framework developed by women of color for women of color. Five years later, a group of Black women gathered in Chicago to develop the reproductive justice framework in response to the Clinton administration's healthcare plan. The Women of African Descent for Reproductive Justice launched the new framework by publishing a full-page statement in the Washington Post critiquing Clinton's plan. The statement was titled "Black Women on

Universal Healthcare Reform," and it included over 800 signatures. Three years later, these women joined 15 other women-of-color-led organizations to form SisterSong and build the reproductive justice movement (Ross et al., 2016).

The reproductive justice framework is built around the idea that Black women's reproductive oppression cannot be separated from the economic and social injustices they experience. Reproductive justice reimagines reproductive rights through the experiences unique to Black women and their communities. The framework recognizes that reproductive freedom cannot be realized without addressing the interlocking systems of oppression, such as poverty, racism, homophobia, police brutality, sexual violence, and pollution that prevent women, especially low-income women and women of color, from controlling their bodies. Reproductive justice emphasizes reproductive control. This includes access to abortion and the right to have a healthy child and raise children in a safe and sustainable environment. The reproductive justice movement shines a light on the reproductive violence Black women have experienced through not only forced birth but also forced sterilization. At the core of the reproductive justice framework is the acknowledgment that the ability to choose does not always equate to agency and liberty if the constraints under which one is presented with choices remain intact and unaddressed. Indeed, full bodily autonomy is not possible if the only choices available are rooted in oppression (Ross & Solinger, 2017).

Reproductive justice intersects with several other social justice frameworks to recognize the barriers to nutritious food, safe housing, a clean, healthy environment, or anything that obstructs the right to full bodily autonomy or the right to freely choose to have a child, to not have a child, or to raise a child in a safe and healthy environment. The concept of intersectionality was first introduced by Kimberlé Crenshaw in 1990 within the context of the reproductive justice framework. Intersectionality seeks to explain how identities based on race and gender intersect within systems of oppression. Intersectionality demands that these identities cannot be separated, as those who embody them cannot separate their experiences tied to each (Crenshaw, 1990). When applying the concept of intersectionality to reproductive freedom, the importance of women's economic, social, and political power to make decisions about their bodies and their families is understood. Accordingly, the environmental impacts of climate change on women and young girls cannot be separated from their reproductive needs and desires. The health and prosperity of a woman's body and fertility are so intrinsically linked to climate that one cannot adequately address the vulnerabilities of one without addressing the burdens of the other. By applying the theory of intersectionality to climate change, one can better understand the relationship between climate and reproductive injustice.

Bodily autonomy is an essential ingredient in reproductive freedom. Where there is reproductive justice, there must be the protection of full bodily autonomy. As researcher Elizabeth Wick writes, "The principle of autonomy is an important foundational concept for the law of human rights, alongside principles of equality and dignity" (Wicks, 2016, p. 6). However, defining what autonomy means is less clear. Traditionally, bodily autonomy is defined as the right to live by one's own law. To achieve bodily autonomy, one must possess both agency and liberty. The concepts of agency, or the ability to make a choice about the body, and liberty, or the ability to freely act based on one's autonomy, have long been debated. The concept of bodily autonomy was first introduced as an expression of self-determination and, at times, isolation from harm. This traditional approach to autonomy centers the quest for full bodily autonomy around personal property and the right to control and protect one's property. Accordingly, the reproductive rights movement is built around the traditional approach to bodily autonomy, or an individual's right to choose how to live. While this may appear to benefit all of society, critiques of the traditional approach point out that individuals are not isolated from those around them and, therefore, the dynamics of relationships should be considered within the concept of autonomy (Wicks, 2016).

In 1989, the same year the National Black Women's Health Project was founded, the idea of relational autonomy was introduced by Jennifer Nedelsky. Rejecting the traditional theory of bodily autonomy rooted in isolation, Nedelsky advocates for a new framework to govern the concept of autonomy with a focus on relationships, interdependencies, and intersectionality. Nedelsky argues that if property is the perfect symbol for the traditional definition of autonomy, then childrearing symbolizes relational autonomy. The relationship of autonomy between parent and child, particularly the ability of the parent to foster or hinder the autonomy and well-being of the child, is the perfect model for relational autonomy. Nedelsky emphasizes the importance of power dynamics within relationships and their impact on individual bodily autonomy (Nedelsky, 1989).

Nedelsky draws parallels from the parent and child relationship to the dynamic between governments and citizens. Relational autonomy illustrates the role of public programs, policymakers, and community organizations in fostering or hindering the autonomy of individuals. Nedelsky's concept of relational autonomy is a facet of reproductive justice. The emphasis on collective action and participation in the democratic process as a central part of autonomy parallels how the underrepresentation of women in positions of power contributes to reproductive injustice (Nedelsky, 1989). Relational autonomy's demand for the collective responsibility of the well-being of all individuals and communities intersects with the reproductive justice framework, particularly the framework's goal to identify the commonality of oppressed women's experiences and to adopt new policies to end that oppression.

The application of autonomy in feminist theory is diverse. However, a few common themes are woven through all the varying viewpoints: anti-individualism, anti-dualism, and the conatus doctrine. All three ideas in the context of autonomy are best explained by borrowing the views of the philosopher Spinoza. Spinoza believed that individuals "are connected by a web of causal relationships governed by natural laws from which [they] cannot escape" (Sharp, 2021, p. 424). In accordance with this theory, the experiences of an individual living in a society cannot be explained in a vacuum and, therefore, autonomy cannot be defined in isolation. In addition, Spinoza's rejection of Cartesian dualism acknowledges the unique experiences of women. The Cartesian doctrine views the body and mind as separate, with an emphasis on the mind for freedom and on the body for servitude. Most feminists reject dualism when defining autonomy due to the irrational annulment of the relationship between the female mind and body, such as the experiences of menstruation, pregnancy, and menopause. Lastly, Spinoza's conatus doctrine acknowledges that feminists seek to understand how patriarchal systems encourage, restrict, and influence individual desires (Sharp, 2021). At the core of this idea is the belief that "the universal striving for self-preservation serves also as a kind of natural pressure upon the state to satisfy the psychic needs of its constituents to belong, to be recognized as worthy of esteem, and thereby to find life livable" (Sharp, 2021, p. 427).

Regulation of Bodily Autonomy

While laws exist to protect autonomy and liberty, particularly in medical law, government regulation of bodily autonomy continues. Current law governs the use of an individual's body in different forms, such as access to abortion care, surrogacy, choices about dying, organ donation, drug use, nudity, and sex work. Regulation of bodily autonomy is motivated by both formal and informal institutions, including patriarchy, capitalism, white supremacy, and purity culture. However, the personhood theory and the no-harm doctrine are the most influential in autonomy regulation. The personhood theory governs who is considered a person with autonomy and when. The no-harm doctrine establishes a standard that bodily autonomy should only be limited to protect against harm to other individuals (Wicks, 2016).

Most recently, the regulation of bodily autonomy has been implicated in conversations around climate change and population control. As a result, the concept of sexual stewards as a solution to climate change has emerged. A sexual steward is "a symbol of the ideal women framed within the logics of private, individual decision-making and choice, who adopts a modicum of embodied environmental responsibility in the service of global development goals, and who helps advocates weave together narratives of the urgency of simultaneously addressing climate change and empowering women around the world to use contraceptives" (Sasser, 2018, p. 4). While population control is not a new concept, advocacy for increasing access to contraceptives to combat climate change, particularly in developing countries, has picked up popularity in recent years. Reminiscent of Norplant and the social policy of the 1990s, which also sought to limit the strain of excess dependents on the economy, pushing birth control on low-income communities, mostly comprised of people of color, reeks of reproductive injustice. Modern population control strategies are false solutions to climate change, focused on "scapegoating those most vulnerable to environmental degradation and climate chaos, displac[ing] attention from root causes by obscuring the role of privileged, hyper-consumers, the fossil fuel industry, and green capitalism/racism in sustaining the current system" (de Onis, 2019, p. 502).

The classification of crimes against the human body or society is often subjective and rooted in historical and cultural context. On one hand, the United States has a history of legislating and litigating restrictions on birth control and abortion cases. On the other hand, compulsory birth control and forced sterilization have been adopted into law by numerous states and upheld by the several courts at varying levels. Further, different social movements, both historical campaigns and modern ones, both advocate for people to choose for themselves whether to have a child and for mandatory birth control to lower carbon emissions and limit undesirable economic consequences. For example, at the same time the reproductive justice movement is rallying for freedom of reproductive choice, the environmental movement is demanding that birth control be thrust upon developing communities to control birth rates and limit lifestyle emissions. Unfortunately, it appears the protection of full bodily autonomy may depend on the decade, the perspective, and the body in question.

References

Crenshaw, K. (1990). Mapping the margins: Intersectionality, identity politics, and violence against women of color. *Stanford Law Review, 43*, 1241.

de Onís, C. M. (2019). Reproductive justice as environmental justice: Contexts, coalitions, and cautions. In *The Routledge companion to motherhood* (pp. 496–509). Routledge.

Licata, J. C. (2020). *The politics of Norplant: Feminism, civil rights, and social policy in the 1990s.* The University of North Carolina at Greensboro.

Nedelsky, J. (1989). Reconceiving autonomy: Sources, thoughts and possibilities. *Yale Journal of Law and Feminism, 1*, 7.

Pumphrey, S. R. (2014). *Reclaiming my body: Black women and the fight for reproductive justice.*

Roberts, D. E. (1999). *Killing the black body: Race, reproduction, and the meaning of liberty.* Vintage.

Ross, L., & Solinger, R. (2017). *Reproductive justice.* University of California Press.

Ross, L., et al. (2016). *Undivided rights: Women of color organizing for reproductive justice.* Haymarket Books.

Sasser, J. S. (2018). *On infertile ground: Population control and women's rights in the era of climate change.* NYU Press.

Sharp, H. (2021). *Spinoza and feminism. A companion to Spinoza* (pp. 422–430).

Wicks, E. (2016). *The state and the body: Legal regulation of bodily autonomy.* Bloomsbury Publishing.

CHAPTER 4

Green Criminology and Ethical Considerations

Central to an individual's right to freely choose whether to have a child and the ability to raise a child in a safe and healthy environment is the health of an individual's surroundings, often measured by the range of surface temperatures, quality of drinking water and food, affordability of the housing market, and access to comprehensive healthcare. Accordingly, the reproductive justice movement finds itself intertwined with the environmental movement. The protection of bodily autonomy and the ability to freely choose not to have a child, to have a child, and to raise a child in a safe and healthy environment demands consideration of one's environment, including access to clean air and clean water, and protection from pollution, toxins, and other harmful substances. Further, many environmental justice teachings can be leveraged when applying the reproductive justice framework, particularly the criminality associated with the inability to have and raise a child in a safe and healthy environment. The environmental movement grounds itself in a few core principles, including the green criminology framework and its three justice-based approaches—environmental justice, species justice, and ecological justice. Importantly, green criminology also addresses the issue of power systems and the role they play in realizing equality, justice, and prosperity. This chapter continues the discussion of the gendered impacts of climate change through the lens of the green criminology framework, a history of environmental justice, and the theory of feminism, and highlights several ethical concerns

S. Haley Knowles, *Diminished Prosperity*,
https://doi.org/10.1007/978-3-032-07068-5_4

that warrant public policy attention at the intersection of full bodily autonomy and the transition to a low-carbon economy.

The term "green criminology" was first introduced by J. Michael Lynch in 1990 as a "variety of class related injustices that maintain an inequitable distribution of power while destroying human life, generating hunger, uprooting and poisoning the environment of all classes, peoples and animals" (Lynch, 1990, p. 3). Since then, green criminology has expanded to include "the causes, consequences, and prevalence of environmental crime and harm, the responses to and prevention of environmental crime and harm by the legal system (civil, criminal, regulatory) and by nongovernmental entities and social movements, as well as the meaning and mediated representations of environmental crime and harm" (Brisman & South, 2019, p. 1). The green criminology framework helps to analyze the impacts of sexism, racism, and capitalism at the intersection of climate justice and reproductive justice and to discuss the ethics and public policy implications of each. These lessons emphasize a feminist reclaiming of full bodily autonomy for poor women and girls of color, given the health and social inequalities they disproportionately experience.

Lessons from Green Criminology

Green criminology is organized into two traditional approaches: the legal-procedural approach, which studies the causes of environmental crime based on violations of law, and the socio-legal approach, which studies environmental harm beyond statutory requirements. While both approaches seek to explain who or what is causing harm, the socio-legal approach expands the framework to address and understand the underlying motives behind why the harm is perpetuated. Therefore, the socio-legal approach provides a suitable methodology to assess racism, sexism, and the exploitation of labor as criminal (Brisman & South, 2019).

Green criminology encompasses a set of philosophies to explain the relationship between humans, nonhumans, and nature. The philosophy of anthropocentrism views humans as superior to nature and nonhumans. Biocentrism believes humans, nonhumans, and nature possess the same worth. Ecocentrism is "based on the idea that humans and their activities are inextricably interconnected with the rest of the natural world" (Brisman & South, 2019, p. 5). The application of these philosophies is accomplished through three justice-based approaches: environmental justice, which demands the equal distribution of the benefits and burdens tied to economic, social, and scientific processes within a society; species justice, which recognizes the utilitarian value and rights of nonhumans; and ecological justice, which "stresses the interrelatedness of life: all living things are bound together and thus all environmental matters are intrinsically transboundary" (Brisman & South, 2019, p. 6).

Lessons applicable to reproductive justice are derived from all three of these approaches. Environmental justice condemns the reality that frontline communities often suffer the worst impacts of climate change, while also being least responsible for the carbon emissions fueling a changing climate. Species justice provides a road map for how classism, often underpinned by racism or sexism, leads to discrimination and prejudice toward groups deemed inferior to others. Ecological justice highlights the importance of intersectionality, the inability to compartmentalize the human experience, and the conflict between commodification and reproductive freedom (Brisman & South, 2019).

Both today and through out history, the impacts of environmental degradation are experienced through race, ethnicity, and gender. For example, during the Civil Rights movement, communities of color fought for civil liberties that spanned not only equal treatment at work, in the voting booth, and on public transit, among others, but also for environmental equality. Environmental injustice was a growing concern of people of color who witnessed their families and neighbors suffering from the burden of toxic waste in their communities. The environmental justice movement, which advocates for the elimination of toxic waste and other hazards associated with industrial or commercial activity unjustly located in low-income communities of color, was launched out of the Civil Rights movement. The first environmental justice action took place in 1968 in Memphis, Tennessee, when Black sanitation workers went on strike to condemn their unsafe and toxic work conditions. However, the term "environmental justice" did not emerge until 1982 when the State of North Carolina decided to locate a toxic waste facility in a small, low-income, predominantly Black community in Warren County, spurring a large community protest (Coolsaet, 2020). Although the protest failed to stop the siting of the toxic waste facility in Warren County, it successfully launched a national conversation on environmental justice. Importantly, the environmental justice movement condemns the disproportionate environmental burdens placed upon historically oppressed communities and demands the equal distribution of the costs of economic progress (Dolšak & Prakash, 2022).

The history of humanity is drenched in the commodification of land, resources, and the human body. This commodification, often paired with colonialism, which frequently is associated with genocide, is rationalized based on perceptions of necessity and economic growth. Yet, the concept of ecocide, which is the destruction of the natural environment through deliberate human behavior, is not universally accepted within the frameworks that govern genocide. An impressive body of literature is emerging to address ecocide and the relationship to "industrial capitalism, in which nature itself is commodified, [which] creates a tension in the pursuit of surplus, resulting in an escalation of consumption and environmental concerns" (Kalkandelen & O'Byrne, 2017, p. 337). The literature showcases how the "interests of post-colonialism and modern corporate imperialism align in asserting that continual economic growth is sustainable and essential" (Crook et al., 2018, p. 310). However, if societies continue to tolerate the environmental and social harm of industrial processes in the name of economic growth, capitalism and consumerism will continue to be two of the biggest drivers of social injustice.

The United States has a long history of accepting environmental degradation in exchange for prosperity. Pollution is so normalized in American culture that in the Fall of 1992, when the Environmental Protection Agency (EPA) sought to respond to public outcry over recent findings that low-income and minority communities are exposed to industrial pollution and toxic waste sites at a much higher rate than affluent and white communities, the EPA created the Office of Environmental Equity to remedy the unequal distribution of pollution and toxic waste. It would take another two years for the EPA to understand the difference between the equal poisoning of all and justice from the poisoning of anyone. While the EPA changed the name of their new office to the Office of Environmental Justice in 1994, the acceptance of environmental harm as an inevitable occurrence that must be distributed demonstrates how deeply capitalism and consumerism are embedded within American society even at the cost of the health and wellbeing of communities (Ewall, 2012).

A Feminist Perspective

In contrast to capitalism, Marxist ecology and ecofeminism reject the notion that continued economic growth is sustainable and essential. Rather, Marxist ecology argues "the dehumanization of human beings…is produced by the capitalist mode of production" (Kalkandelen & O'Byrne, 2017, p. 337). Marxist ecology believes that the ecological crisis and social injustice are interconnected, and ecofeminists agree, paralleling sexism with environmental degradation through the analysis of society's exploitation of both women and nature. Ecofeminism was introduced in 1974 by Françoise d'Eaubonne and calls for a collaborative society, with no dominant group (Zein & Setiawan, 2017). Ecofeminists explore how the institutions of race, gender, and power contribute, individually and collectively, to both the oppression of women and the destruction of nature. Like Marxist ecology, ecofeminists argue that the human body cannot be separated from nature, creating an axiom at the intersection of women's oppression and a dying planet (Kalkandelen & O'Byrne, 2017). Ecofeminists further highlight the importance of women as decision-makers in decoupling carbon emissions from economic growth (Ergas & York, 2012).

The relationship between gender inequality and environmental degradation is well-supported by the literature. A 2020 fixed-effects panel regression model assessed the relationship between carbon emissions and economic growth at varying degrees of gender inequality. The study found that "...increasing gender inequality in both developed countries and less developed countries leads to a tighter coupling between economic growth and CO_2 emissions [but] the association of gender inequality on the relationship between economic development and CO_2 emissions is much stronger in less developed countries" (McGee et al., 2020, p. 263). Building upon previous research that inversely correlates women's political status and carbon emissions, the literature concludes that increasing gender equality may lead to curbing climate change (Ergas & York, 2012). These findings raise the question of who is best suited to address and remedy incidents of injustice.

Assessing criminology through a feminist lens parallels many of the concepts previously discussed. For example, feminist criminology is "a paradigm that studies and explains criminal offending and victimization as well as institutionalized responses to these problems as fundamentally gendered" (Renzetti, 2018, p. 76). At the root of feminist criminology is the acknowledgment that the experiences, behaviors, and desires of women have largely been excluded from traditional approaches to criminology, mainly due to patriarchy. Further, Marxist feminism, as opposed to liberal feminism, emphasizes the role of social class and, therefore, the role of capitalism, in the oppression of women (Renzetti, 2018). These lessons from green criminology and feminist theory demand a feminist reclaiming of full bodily autonomy for low-income women and women of color through the reevaluation of positions of power.

Ethical Considerations

The public policy implications of climate change, environmental racism, sexism, and the exploitation of laborers are complicated; consensus is not often reached. However, the frameworks discussed in this chapter propose a series of solutions to lessen and remedy the social injustices connected to each. From green criminology, policy solutions must begin with "learning from—or endeavoring to understand—previous mistakes and omissions, as well as anticipating future risks and threats" (Brisman & South, 2019, p. 3). Further, "a green criminology concerned with the creation of effective systems for the administration of environmental justice should be supporting a proposal to introduce into international law new measures to address contemporary ecocidal trends" (Crook et al., 2018, p. 302). From an ecofeminism perspective, remedy resides in a power shift to provide better representation within the policymaking process of those most impacted by social injustice. "Women play a vital role in the societal response to climate change; their participation at all levels has been shown to result in greater responsiveness to citizens' needs and often increases in cooperation across party and ethnic lines, generally resulting in sustainable outcomes" (Sorensen et al., 2018, p. 5).

The lessons from green criminology, ecofeminism, and relational autonomy combine to strengthen the pursuit of full bodily autonomy. However, the conclusions drawn from each are rooted in oppressive systems. Green criminology ignores the unique experiences of women and builds its theory upon mostly a man's perception. Ecofeminism and relational autonomy embrace the vital role gender and social class play in oppression but largely ignore the role of race. Ecofeminists advocate for increased access to contraceptives, but the notion that all women are liberated through population control is based on the perspective of affluent white women. Nedelsky's (1989) assertion that bodily autonomy is best achieved through stronger relationships and collaboration is driven by someone who does not experience systemic racism present in a society. While reproductive justice revolves around the theory of intersectionality, the importance of relational decision-making often gets lost in the application of the framework, focusing on the right to achieve reproductive freedom but without a clear standard to determine when (and by whom) those rights are fulfilled.

While a rich and complex body of literature evidences the detriment to women and low-income communities from a changing climate, the knowledge of this harm does not motivate policymakers to act. With most policymakers as white affluent men who are not substantially impacted by climate change, those making policy decisions do not understand the unique needs of those most impacted by a warming planet or appreciate the urgency to act. As such, the power dynamics within formal and informal institutions sustain the barriers to full bodily autonomy for individuals with lower socioeconomic status. Further, remediation of this injustice through the traditional approach to autonomy falls short. With the traditional approach to autonomy focused on self-determination and protecting one's property, policymakers are persuaded to protect their own property to preserve their own autonomy in isolation. As a result, sustained gender inequality not only perpetuates oppression of marginalized communities and environmental degradation, but it also protects the bodily autonomy of those in power.

To truly achieve full bodily autonomy, individuals must be able to decide what that means for their body. Missing from the reproductive justice framework is the following tenet: *The right to decide when full bodily autonomy is achieved depends on and is derived from one's lived experience.* While the lessons of the previously discussed frameworks bring us closer to full bodily autonomy, the lived experiences of those most marginalized must be prioritized to realize reproductive justice. Indeed, absent a similarly lived and shared experience, it is not possible for others to genuinely understand the unique needs of those who experience differently, nor is it ethical (e.g., equitable, good) for others to decide how to remedy an injustice that does not impact them (Young, 2011). Without the right to decide when full bodily autonomy is achieved based on group differentiated experiences, the doctrine of full bodily autonomy will continue to be exploited by those in power to protect their own property and disregard the most vulnerable. The refusal to put forth effective policy solutions to solve and mitigate climate change impacts—given those who are most harmed by these effects—is an ethical dilemma.

Full bodily autonomy and the right to choose to have a child, to not have a child, and to raise a child in a safe and healthy environment is vital for equitable communities. Preventing and mitigating the impacts of climate is important for healthy communities. In confronting climate change and reproductive injustice, policy solutions must consider the interdependencies and intersections between the two to address the root causes of both. The refusal to tackle the causes of climate change is, in part, due to who is most impacted by climate change. Women have historically experienced oppression, particularly within the communities on the frontlines of climate change, along with having their unique healthcare needs ignored when compared to the population at large. Several examples exist of this phenomenon, from safety testing of automobiles that ignore the smaller stature of women to the development of vaccines that ignore interactions with birth control. Thus, the reclaiming of bodily autonomy through the lived experiences of women and communities of color is essential for reproductive justice to be realized. And, the reclamation begins with understanding, dissecting, and reallocating informal and formal systems of power.

References

Brisman, A., & South, N. (2019). Green criminology and environmental crimes and harms. *Sociology Compass, 13*(1), e12650.

Coolsaet, B. (2020). *Environmental justice: Key issues.* Routledge.

Crook, M., et al. (2018). Ecocide, genocide, capitalism, and colonialism: Consequences for indigenous peoples and global ecosystems environments. *Theoretical Criminology, 22*(3), 298–317.

Dolšak, N., & Prakash, A. (2022). Three faces of climate justice. *Annual Review of Political Science, 25*, 283–301.

Ergas, C., & York, R. (2012). Women's status and carbon dioxide emissions: A quantitative cross-national analysis. *Social Science Research, 41*(4), 965–976.

Ewall, M. (2012). Legal tools for environmental equity vs. environmental justice. *Sustainable Development Law & Policy, 13*, 4.

Kalkandelen, K., & O'Byrne, D. (2017). On ecocide: Toward a conceptual framework. *Distinktion: Journal of Social Theory, 18*(3), 333–349.

Lynch, M. J. (1990). The greening of criminology: A perspective for the 1990s. Reprinted in N. South & P. Beirne (Eds.). In *Green criminology.*

McGee, J. A., et al. (2020). Gender inequality, reproductive justice, and decoupling economic growth and emissions: A panel analysis of the moderating association of gender equality on the relationship between economic growth and CO2 emissions. *Environmental Sociology, 6*(3), 254–267.

Nedelsky, J. (1989). Reconceiving autonomy: Sources, thoughts and possibilities. *Yale Journal of Law and Feminism, 1*, 7.

Renzetti, C. M. (2018). Feminist perspectives. In *Routledge handbook of critical criminology* (pp. 74–82). Routledge.

Sorensen, C., et al. (2018). Climate change and women's health: Impacts and policy directions. *PLoS Medicine, 15*(7), e1002603.

Young, I. (2011). *Justice and the politics of difference.* Princeton University Press.

Zein, L. F., & Setiawan, A. R. (2017). *General overview of ecofeminism.* OSF Preprints.

PART II

Critiques of White Feminism, Corporate Colonialism, and the Rise of "Girlbosses": Acknowledging Who Is Most Responsible for Climate Change

CHAPTER 5

A History of Settler Colonialism

The distribution of the impacts of climate change inversely correlates with the distribution of power. Specifically, the less power an individual or community has, the more impacted they are by a changing climate, such as the consequences of increased surface temperatures and incidents of extreme weather. Conversely, the more power an individual or community has, the less impacted they are by the consequences of climate change, often due to those with higher socioeconomic status being able to relocate following an extreme weather event and afford higher prices of food, utilities, healthcare, and housing. Importantly, the foundation for power distribution among systems, institutions, and communities was laid hundreds of years ago by colonial agents, with remnants of their power dynamics persisting today.

Part II of this book evaluates who is most responsible for climate change and examines some of the early incidents of white feminism and corporate colonialism, and the resulting rise of corporate white feminists as high-paid influencers who contribute, although sometimes inadvertently, to the exploitation and ongoing colonialism of Indigenous women. By recalling the historical legacy of settler colonialism, especially the exploitation of Indigenous women and their ongoing oppression, and by examining the role of corporate expansion and consumerism, the next four chapters aim to demonstrate how white feminists create and maintain corporate colonialism, and ultimately the colonization of Indigenous and other non-white women.

S. Haley Knowles, *Diminished Prosperity*, https://doi.org/10.1007/978-3-032-07068-5_5

The Criminalization of Indigenous Culture

The ideologies of white supremacy, patriarchy, and capitalism pave the way for settlers to gain control over territories, nations, and people, founded on the perceived superiority of the white way of life and the expected financial benefits of investments in other countries. Colonialism can be divided into two broad categories: external colonialism and internal colonialism, with the former being a method of resource extraction for chattel and the latter serving to police and manage land, people, and everything indigenous for the purpose of dominion by and assimilation to the white elite (Tuck & Yang, 2021). Settler colonialism throughout North America accomplished both goals, along with establishing the concept of prosperity within a Christian, capitalist culture.

Throughout the history of North America, prosperity, as defined in the Western context, has often been tied to the othering of cultures and peoples that threaten capitalism. As anthropologist James Morris Blaut, a staunch critic of eurocentrism, points out, "[It was obvious to many] that a Christian god would guide his people to civilization and modernity—and thus of course to capitalism" (1989, p. 260). While early colonizers believed that progress would come to non-European people only through the diffusion of European and Christian ideals, colonialism was inherently about the accumulation of capital. From gold and silver to leather, cloth, and spices, colonialism revolved around the flow of goods and labor, including the enslavement of Indigenous Peoples throughout the Americas (Blaut, 1989). Thus, the history of capitalism can be traced back to the domination and exploitation of nature and cultures, particularly Indigenous women. This quest for wealth was often accompanied by rationalizations based on missionary work and driven by the Christian concepts of morality.

Beginning in the 1500s and continuing for more than four centuries, colonial agents in what is now the United States and Canada convinced themselves and others that their mission was righteous by presenting Indigenous women as people who were either fragile and needed saving or who were violent and needed controlling. Colonial agents portrayed Indigenous women as inferior, uneducated, unfit, and untamed. Not only did European colonizers bring with them their agenda to rape and pillage land, but they also overwhelmed tribal nations with their demand for female subjugation, forcing Indigenous women to fight to maintain their identities, which were often rooted in matriarchal tradition (Stevenson, 1999). Further, settler colonialism brought fur traders and Christian missionaries, both of whom were shocked when interacting with Indigenous women. Colonial agents viewed Native peoples' diverse practices of gender and sexuality as primitivity (Morgensen, 2010). They were used to the submissive, meek qualities of European women, not the strong, fierce qualities of many Indigenous women they encountered. The resilient, independent nature of Indigenous women fueled the need for colonial agents to develop a strategy to dominate, control, and invalidate them.

As a result, colonial agents organized Native women into two broad stereotypes: an "Indian Princess" or a "Squaw Drudge," both of whom needed rescuing (Stevenson, 1999, p. 57). Further, patriarchy, which often assigned to women feminine, domestic duties and encouraged them to pursue outlets within feminine creativity, such as painting, dancing, and knitting, was thrust upon Indigenous women by colonizers. This led to colonized Indigenous women being excluded from what colonial agents deemed masculine duties, such as voting or participating in decisions related to the management of their land. Many Native women lost the freedom and autonomy they were accustomed to, including the right to own property or initiate divorce proceedings (Stevenson, 1999). Indigenous women, many of whom were once highly respected leaders in their communities, were stripped of their power and status through confinement and reeducation.

During the second half of the 1700s, Spanish missionaries and priests arrived in what is now California with a mission to "denaturalize" and "humanize" Indigenous Peoples, which they referred to as savages (Teran, 2016, p. 20). In particular, young girls and women were captured and taken from their homes by Spanish soldiers and held in monjeríos, which were large buildings similar to prisons that served as "a site of gendering and racialization" (Teran, 2016, p. 21). Native women were routinely tortured, shamed, and sexually assaulted at the hands of the Catholic Church in an attempt to control their sexuality and perceived immorality. Further, beginning in the 1700s and continuing into the twentieth century, the United States established more than 500 government-run Indian boarding schools across most of the Western United States (Pegoraro, 2015). The schools were used to reeducate and assimilate Indigenous children to remove all signs of their culture, along with punishing those who held onto their customs, beliefs, or language, including any parent who refused to send their children to the boarding schools (Ross, 2016).

Stories of crimes and horrors suffered at these residential schools are vast, from rape and torture to forced sterilization and death (Pegoraro, 2015).

The missions and government-run boarding schools served as some of the nation's first prisons by assigning deviance to nonwhite, non-Christian behaviors and controlling the movement of nonwhite, non-Christian populations. Boarding schools and other federal policies, such as the pass system, which required anyone leaving the reservation to receive a pass from the Indian agent and criminalized those caught without one, controlled both the physical and cultural movement of Indigenous Nations across North America. This resulted in not only Indigenous Peoples being contained but also removed from their land, which, when left unattended, increased opportunities for the land to be seized by settlers. The beginning of the disproportionate incarceration of Indigenous Peoples in the United States and Canada can be traced back to this era of criminalization and confinement of Indigenous culture and ideas. As sociologist Luana Ross (2016), a member of the Confederated Salish and Kootenai Tribes, explains, "a thorough understanding of Native criminality requires tracing the evolution of federal policies, the process of being made deviant, and the politics of confinement" (p. 3).

Indian agents, first introduced by the Federal Nonintercourse Act of 1793, were used by the United States government to facilitate the federal regulation of Indian affairs. For example, the law required Indian agents to oversee land sales by or from Indians, along with guiding First Nations in their assimilation into American society by changing their agricultural practices, counseling them on their management of money, and transforming their household domestic activities. Over the next 100 years, the role evolved to a responsibility for distributing federal payments to tribes based on negotiated treaties, ensuring cooperation between First Nations and the military, and overseeing the removal of Indigenous Peoples from land at the request of the federal government (Prucha, 1995). The impact of these federal policies on First Nations is fiercely debated among historians, with American historian Francis Paul Prucha painting perhaps the most favorable view of government officials rejecting the calls from frontiersmen to exterminate Indians as a form of paternalism, or what he calls "a determination to do what was best for the Indians according to white norms, which translated into protection, subsistence of the destitute, punishment of the unruly, and eventually taking Indians by the hand and leading them along the path to white civilization and Christianity" (1995, p. 27). Regardless of how historians spin it, by the 1830s, the primary role of Indian agents was to supervise commercial transactions between Indians and traders, which led to Indians losing the ability to make decisions over their land, assets, and, ultimately, their own autonomy, and cultivating the institutionalized control of First Nations' economic and social status across the United States.

The Displacement and Erasure of Indigenous Lifeways

In 1862, the largest mass execution in the history of the United States took place during the Civil War at the order of President Abraham Lincoln. The order, which sanctioned the execution of 38 Dakota men, was a cruel conclusion to the Dakota War and sparked three decades of Indian wars between the First Nation tribes and the United States military (Martinez, 2013). However, the underlying cause of the conflicts began over ten years prior when, beginning in 1850, the culture, customs, and lifeways of First Nations began to decline due to being pushed off their native land and forced to assimilate to the white settler state (Tallbear, 2016). When promises made to First Nations for control of their land by the American government went unfulfilled, treaties were once again negotiated. The new treaties, driven by the demand of frontiersmen and policymakers for tribes to make way for economic expansion and progress, reduced the Dakota's land by half, decreasing their hunting grounds significantly and resulting in many of the Dakota going hungry. The decreasing quality of life of the Dakota ultimately led to armed conflict with settlers, over 300 Dakota men sentenced to death for their participation in the war, more than 1600 women, children, and elderly Dakota confined to concentration camps, with hundreds more dying at the camps from illness and disease, and a cry by the Governor for all Indians in what is today Minnesota to be exterminated (Martinez, 2013).

By 1863, the United States passed a law to annul all treaties with the Dakota and take control of all their land. Shortly after, a second law was passed to expel not only the Dakota from Minnesota and their ancestral homelands, but all Indigenous Peoples, including the Ho-Chunk, who had not participated in the Dakota War but whose land was deemed by colonial agents to be pristine agricultural terrain (Martinez, 2013). By 1880, Indian agents were instructed to enforce a number of laws focused on the economic and social assimilation of First Nations, such as the prohibition of liquor, the supervision of the instruction of English and the reeducation of Indian children, ensuring that jobs are not "given to white men which can be done by Indians," and prohibiting Indians from leaving their respective reservations without a permit (Prucha, 1995, p. 293). Less than four years later, beginning in 1881, Indian agents started compiling annual reports on each of the Indigenous Peoples living on the reservations they oversaw, including name, gender, and family status, among other statistical information (NA, 2016).

Toward the end of the nineteenth century, following decades of violence-based control and removal tactics by white settlers and Indian agents, the Native American Ghost Dance ritual began to spread like wildfire throughout much of the Western United States. Native Americans saw the dance as a way to honor their heritage and reunite with spirits of the dead and, as a result, conjure renewed power to end American westward expansion. The dance was originally practiced by the Nevada Northern Paiute and promoted by their shaman, Wovoka, or Jack Wilson, who claimed that the proper practice of the dance for five consecutive days in the appropriate intervals would end colonial expansion (Mooney, 1991). Through the Ghost Dance, many Indigenous Peoples literally believed they could dance a new world into existence. Others found comfort, strength, and solace by taking part in the dance ritual. Almost all believed the goal of the ritual was to promote unity, coexistence, and peace. To settlers, however, the dance was seen as a dangerous form of collaboration among different tribal nations, which frightened the United States government and led to concerns over a possible coordinated uprising of Indigenous Peoples across multiple tribes in various geographical regions. By 1890, Indian agents banned all Indigenous Peoples from practicing the Ghost Dance.

In the Fall of 1890, after the United States government once again violated a Lakota treaty by significantly reducing the Great Sioux Reservation in South Dakota to accommodate white settlers on a quest for gold and Lakota families were forced to relocate to homesteads to farm and eventually starve after their harvests failed in the semi-arid region, the Lakota, who added their own interpretation to the dance ritual, turned to the Ghost Dance for resistance and the removal of non-Indians from their land. By December of 1890, the United States Army deployed to the Lakota reservation to arrest Chief Sitting Bull for failing to stop his people from practicing the Ghost Dance. What followed was a series of events, including the death of Chief Sitting Bull, armed conflict between the Sioux Nation and the United States military, and the killing of up to 300 Lakota, mostly women and children, which ultimately led to the Lakota officially surrendering to the United States Army (Mooney, 1991).

The 1890 surrender of the Lakota brought an end to the bloody Indian wars, which spanned three decades and hundreds of miles across the Great

Plains, from the Dakota War of 1862, the 1863 massacre at Bear River in Idaho, and the 1864 massacre at Sand Creek in Colorado to the 1890 massacre at Wounded Knee in South Dakota (Martinez, 2013). The Indian wars, often overshadowed in history books by tales from the Civil War, which coincidentally took place around the same time and conveniently, but erroneously, paint the North as the more humane side, were the primary driver of the sharp decrease in Indian populations at the turn of the century. By 1900, the population of Indigenous Peoples in the United States, estimated to be about 7 million before settlers arrived, had plummeted to around 250,000 (Thornton, 1987). However, the criminalization, imprisonment, and killing of First Nations by the United States government went beyond physical constraint to control not only the physical movement of Indigenous Peoples but also oppress their languages, customs, and spiritual practices.

Settler colonialism not only physically removed and displaced tribal nations from their land, but also forced Native cultures underground as a way to erase their heritage. As a result, many Native women (and men) became afraid of teaching their children their language, customs, and rituals for fear of violence from colonial agents. The Nez Percé tribe, who have lived on the Columbia River Plateau for centuries, have a word for imprisonment that literally means darkness, illustrating how for the Nez Percé, and many other tribal nations, imprisonment is not a place, it is an ongoing condition (Ross, 2016). Further, the colonial discourse that continues into modern times often "conflates racialized terrorists with sexual perversion [to] uphold the heteronormativity of white citizens as in need of enhancement or defense," which only further sustains the settler state as perceived protection against immorality or violent uprising (Morgensen, 2010, p. 105). Therefore, the confinement and assault of Indigenous Peoples through federal policy that results in the criminalization and incarceration of Indigenous lifeways continues today as a form of ongoing settler colonialism, not only through controlling the physical movement of tribal nations, but also through a mental, spiritual, and emotional condition of oppression, exploitation, and ultimately, erasure.

Weaponizing Public Policy to Control First Nations

The rapid population decline of Indigenous Peoples that began in the 1700s and continued throughout the 1800s can be traced to a number of historical events, from violent colonial practices and the spread of infectious disease to government interference, isolation, and forced sterilization. For more than four centuries, Indigenous Peoples were victims of not only war, violence, and removal but were also often stripped of their ability to reproduce. For example, in the United States, the California missions and Indian boarding schools saw a sharp decline in fertility rates among the Native women housed there due to the poor living conditions, the spread of infectious disease introduced by their European educators, and a lack of adequate medical treatment (Teran, 2016). In addition, continuing throughout the twentieth century and throughout North America, Native women, of whom were both ill and healthy, were often confined to government hospitals based on their Indian status, where they experienced unimaginable violence and inadequate health services, often resulting in their infertility or death.

For example, in Canada, the Department of Indian Affairs forced many Indigenous women into Indian hospitals not based on disease or illness, but on their Indian status in an attempt to control and assimilate them. Many stories of healthy First Nation women being used to test experimental drugs and procedures, along with undergoing electroshock therapy and sterilization without their consent or knowledge, emerged during the second half of the 1900s. In addition, Indigenous women who were battling an illness did not receive proper care at government hospitals but were also not allowed to leave. Further, in 1953, the Canadian Indian Act was amended to make it a crime for an Indigenous individual to refuse to see a doctor or leave a hospital before being officially discharged (Lux, 2016). As a result, Indigenous women not only experienced unimaginable trauma and terror at the hands of colonial agents disguised as medical care, including being beaten or shocked to death or left in isolation to die, but also faced yet another mode of criminalization for merely existing.

Like Canada, the United States leaned on federal policy to assist with the management of tribal populations. During the second half of the twentieth century, Indigenous populations began to rebound. In 1970, the United States Census showed that the average Native American woman bore 3.79 children compared to 1.79 children for all other groups in the United States (Pegoraro, 2015). Soon after, the Population Research Act of 1970 was passed and led to Indigenous women throughout the United States being targeted for birth control and sterilization, often without their knowledge or consent (Carpio, 2004). Following centuries of violent removal, displacement, and relocation, many Indigenous Nations struggled with poverty and looked to Indian Health Services (IHS) for healthcare. However, medical records show that IHS often abused authoritative positions of power and language barriers to manipulate, force, and deceive Indigenous women into sterilization procedures, as the government viewed birth control and sterilization as methods of controlling government dependency. In 1979, Lehman Brightman, a member of the Great Sioux Nation and founder of United Native Americans, found that 42% of Native American women, equating to around 60,000 to 70,000 women, had been coercively sterilized by physicians in partnership with IHS. As a result, by 1980, the average number of children born to Native American women dropped to 1.3 (Pegoraro, 2015).

As previously discussed, the widespread reproductive injustices experienced by Indigenous women were exacerbated by white women and the reproductive rights movement, who advocated for women's sexual liberation and increased access to contraceptives, along with policymakers and eugenicists who advocated for policies that tied compulsory birth control to welfare benefits in an effort to decrease birth rates amongst low-income communities of color. Birth control advocates gained momentum not only for advocating for white women's ability to control their fertility but also for impoverished mothers to stop reproducing. While forced sterilization and social policy of the 1900s may be criticized today as a form of reproductive coercion to discourage poor communities from reproducing, many modern progressives and economists continue to push birth control and the regulation of fertility as a method to address poverty, improve the welfare of humanity, and advance economic progress for the white elite.

Today, white settler states continue to rationalize the confinement and subjugation of Indigenous women. Further, while the history of Spanish missions and government boarding schools is largely condemned and considered an injustice of the past, Indigenous women continue to resist the reeducation of their daughters by the settler state. For example, through Indigenous feminism, many First Nations contain matrilineal power structures with alternate concepts of gender identity and family roles; however, the patriarchal notion of womanhood continues to dominate most societies throughout North America and Christian values are largely considered the norm by Western societies, which continue to dictate the modern understanding of morality. All these ideas and concepts, which directly contradict many of the traditions and beliefs of tribal nations, drive continued settler colonialism and the forceful confinement, sterilization, and reeducation of Indigenous women and their children as an ongoing structure, not an isolated event of the past. A throughline exists between the historical exploitation of Indigenous Peoples and land and the ongoing modern dominion over human and natural capital that continues to adversely impact society based on the relationship between increasing climate change impacts and decreasing birth rates.

References

Blaut, J. M. (1989). Colonialism and the Rise of Capitalism. *Science & Society, 53*(3), 260–296.

Carpio, M. V. (2004). The lost generation: American Indian women and sterilization abuse. *Social Justice, 31*(4 (98)), 40–53.

Lux, M. K. (2016). *Separate beds: a history of Indian hospitals in Canada, 1920s–1980s*. University of Toronto Press.

Martínez, D. (2013). Remembering the thirty-eight: Abraham Lincoln, the Dakota, and the US war on barbarism. *Wicazo Sa Review, 28*(2), 5–29.

Mooney, J. (1991). *The ghost-dance religion and the Sioux outbreak of 1890* (Vol. 176). U of Nebraska Press.

Morgensen, S. L. (2010). Settler homonationalism: Theorizing settler colonialism within queer modernities. *GLQ: A Journal of Lesbian and Gay Studies, 16*(1-2), 105–131.

National Archives. (2016). *Indian Census roles, 1885–1940*. Legal and Administrative Background: The U.S. National Archives and Records Administration.

Pegoraro, L. (2015). Second-rate victims: The forced sterilization of Indigenous peoples in the USA and Canada. *Settler Colonial Studies, 5*(2), 161–173.

Prucha, F. P. (1995). *The great father: The United States government and the American Indians* (Vol. 1). U of Nebraska Press.

Ross, L. (Ed.). (2016). *Settler colonialism and the legislating of criminality*. American Indian Studies Center, University of California.

Ross, L., et al. (2016). *Undivided rights: Women of color organizing for reproductive justice*. Haymarket Books.

Stevenson, W. (1999). Colonialism and first nations women in Canada. In *Scratching the surface: Canadian anti-racist feminist thought* (pp. 49–80).

TallBear, K. (2016). Annual meeting: The US-Dakota War and failed settler kinship. *Anthropology News, 57*(9), e92–e95.

Teran, J. (2016). The violent legacies of the California missions: Mapping the origins of Native women's mass incarceration. *American Indian Culture and Research Journal, 40*(1), 19–32.

Thornton, R. (1987). *American Indian holocaust and survival: A population history since 1492* (Vol. 186). University of Oklahoma Press.

Tuck, E., & Yang, K. W. (2021). Decolonization is not a metaphor. *Tabula Rasa*, (38), 61–111.

CHAPTER 6

Climate Change and Declining Birth Rates

A common misconception of one of the tenets of reproductive justice is that if an individual is freely able to engage in family planning, the individual has full bodily autonomy. However, to freely choose to have or not to have a child, an individual must be able to make that choice free of any influence from or restriction by other factors, such as the cost of raising kids. If an individual desires to have a child but chooses not to over concerns about being able to afford the cost of raising the child, then the individual has not freely chosen and, therefore, does not have full bodily autonomy. With concerns over the economy and the cost of living rising year over year, birth rates are seeing a steady decline across geographies and demographics. Could climate change and the resulting rise in living costs be to blame for the global drop in birth rates? This chapter explores how dominion over land and resources continues to adversely impact society by discussing the relationship between climate change and birth rates.

Every two years, the Organisation for Economic Co-operation and Development (OECD) publishes its *Society at a Glance* report to provide internationally comparable data on demography and family characteristics, employment and wealth, mobility, and housing. The 2024 edition, which includes 25 social indicators and data from 38 OECD member countries, reported that, in 2022, the global fertility rate dropped from 3.3 children per woman in 1960 to 1.5 children per woman, which is below the replacement rate of 2.1. A common metric for tracking birth and fertility rates is

S. Haley Knowles, *Diminished Prosperity*,
https://doi.org/10.1007/978-3-032-07068-5_6

the replacement rate, or the number of births needed to maintain current population levels. The drop is surprising to most, partly because it includes women who differ by race, ethnicity, age, income, labor participation, and education, but especially since national birth rates are reporting around 10–20% below estimates from the United Nations (Ip & Adamy, 2024). However, what is more puzzling is that no one can quite figure out the primary driver of the decline in birth rates.

Model 4: The Relationship Between Cost of Living and Birth Rates

To better understand how the rising costs of food, housing, and healthcare contribute to reproductive injustice, I explored how three components within the cost of raising a child, measured by food, housing, and healthcare expenditures, impact one component of full bodily autonomy, measured by birth rates. Leveraging the results of the first three linear fixed effects regression models discussed in Chap. 2, in model 4, I employ the weighted average of all three dependent variables, which represent food, housing, and healthcare expenditures, to explore the link between the cost of living and birth rates, measured by the total number of births organized by year and state. Model 4 is controlled for fertility based on state-level fertility rates from the National Center for Health Statistics (NCHS), part of the Center for Disease Control and Prevention (CDCP), which tracks the percentage of females aged 15–44 in the current population. Based on the literature, increased costs of raising a child can result in restricting the ability to freely choose to have a child and, ultimately, increased barriers to full bodily autonomy represented by declining birth rates. Table 6.1 summarizes model 4's hypothesis and variables.

The independent variable for model 4 is the annual dollar amount of food, housing, and healthcare expenditures, as defined and calculated by models 1, 2, and 3 using the BEA CPE Index by State, measured in millions of current dollars at the state level, organized by year for the years 1997 to 2021, adjusted for inflation using the CPI, and divided by the corresponding state population based on data from FRED to generate an average cost of living.

Figure 6.1 shows the state-level difference for annual rate of living costs, measured in millions of dollars and divided by the state population, for the years 1997 through 2021. Across the board, the cost of living has significantly increased for all states within the contiguous United States

Table 6.1 Model 4 hypothesis and data

Hypothesis 4	*Independent variable*	*Dependent variable*	*Controls*
H4: As the cost of raising a child increases, the number of children born will decrease.	**IV4:** Average annual cost of living	**DV4:** Annual number of births	1. Economic recession 2. Fertility rate
Variable description	Annual cost of living, or the average cost of food, housing, and healthcare expenditures at the state level, divided by the population	Annual number of births at the state level, divided by the population	1. Number of quarters spent in economic recession at the national level for each calendar year on a scale of 0 to 4 2. Percentage of females aged 15 to 44 in the population at the state level

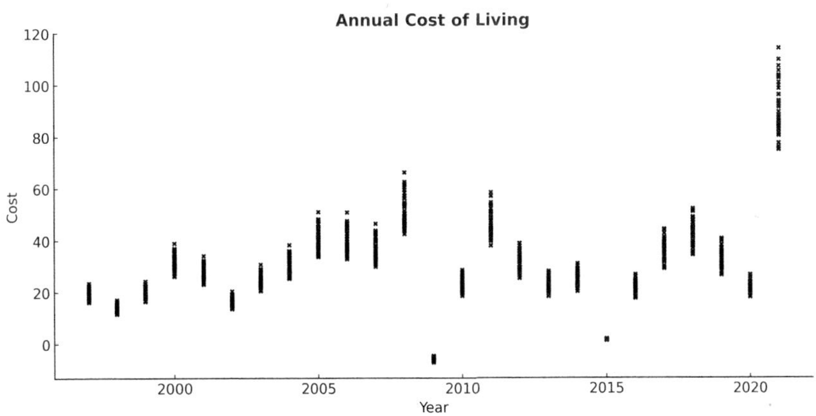

Fig. 6.1 Annual cost of living

over the last 25 years, with variance across the states increasing from 3.72 in 1997 to 99.2 in 2021. The largest increase in the cost of living occurred in 2021 across the majority of the contiguous United States, with living costs being highest in the States of Massachusetts, New Hampshire, Vermont, California, New York, Connecticut, Washington, Delaware, and Colorado, at an average of $105,780 per capita. In contrast, the States of Alabama, Utah, Arkansas, Oklahoma, and Mississippi had the lowest cost of living at an average of $75,570 per capita.

Figures 6.2 and 6.3 are a set of two maps of the continental United States—one for the year 1997 and one for the year 2021—representing the state-level difference for annual cost of living with a darker red indicating higher costs. Figure 6.4 illustrates the percentage change of the state-level difference for the annual cost of living by US state for the years 1997–2021.

In model 4, the dependent variable is the annual number of births for each state, calculated based on data from the CDCP, organized by year for the years 1997 to 2021, and divided by the corresponding state population based on data from FRED to generate a state-level birth rate. The model controls for the fertility rate, based on the percentage of females aged 15 to 44 in the population at the state level, per the Fertility Rates by State dataset from the CDCP, and is organized by year for the years 1997 to 2021.

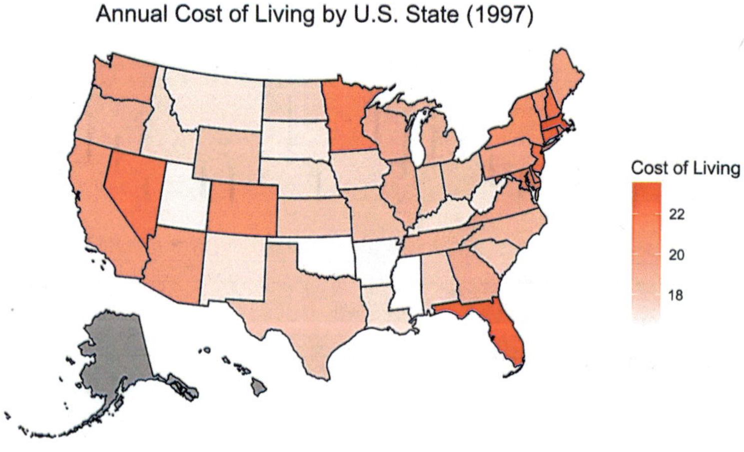

Fig. 6.2 Annual cost of living in 1997

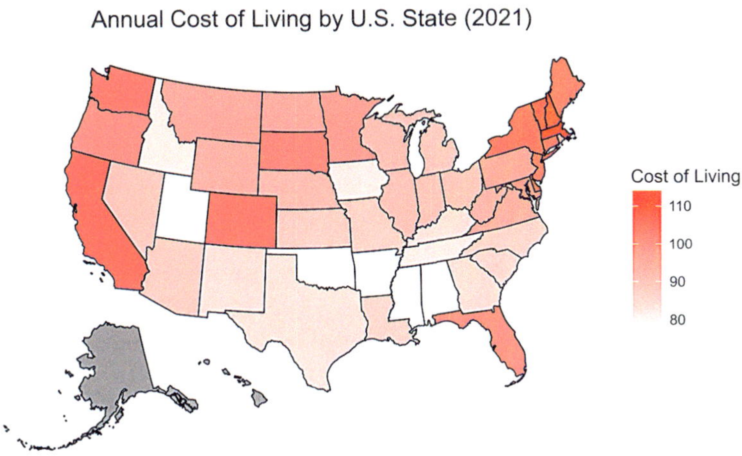

Fig. 6.3 Annual cost of living in 2021

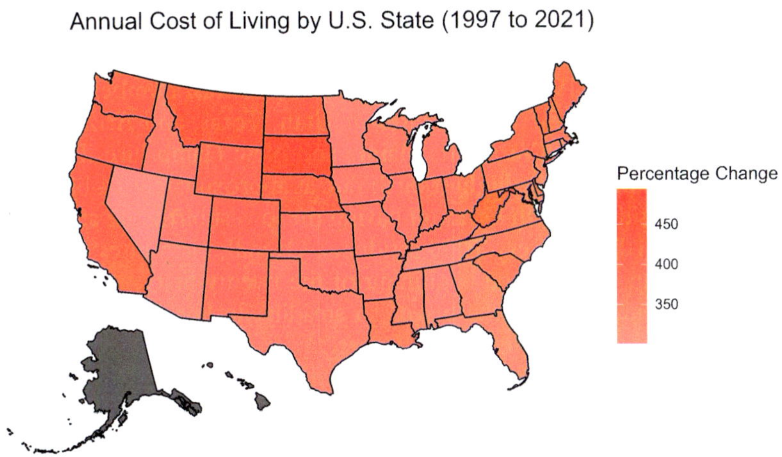

Fig. 6.4 Change in annual cost of living

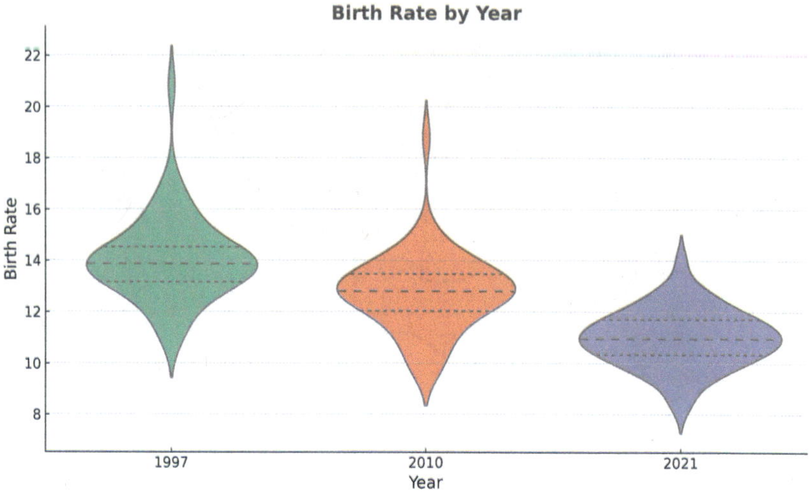

Fig. 6.5 Annual birth rate

Figure 6.5 shows the state-level difference in the annual number of births for the years 1997, 2010, and 2021. The annual number of births continues to be highest in the States of Utah, Texas, and Arizona, and lowest in the States of Vermont, Maine, and New Hampshire. In 2021, the State of Utah had the highest birth rate at approximately 14 births for every 1000 residents, with Vermont having the lowest birth rate at approximately 8.3 births for every 1000 residents. Overall, birth rates remained fairly steady up through the beginning of the twenty-first century, when the annual number of births began to steadily decline across all states throughout the contiguous United States.

Figures 6.6 and 6.7 show the state-level difference in annual birth rates for the year 1997 versus 2021. The darker the color of green, the higher the rate of birth. The lightest color of green indicates the lowest rate of birth. Figure 6.8 illustrates the percentage change of the state-level difference for annual rate of births by US state for the years 1997–2021. The darker the color of red, the higher the percentage of change.

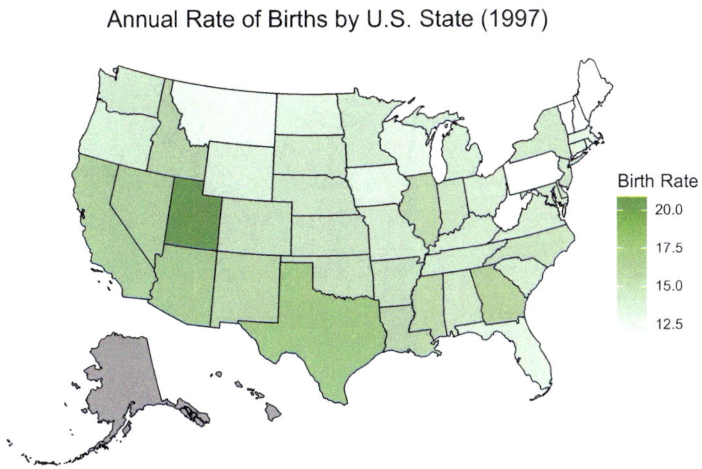

Fig. 6.6 Annual birth rate in 1997

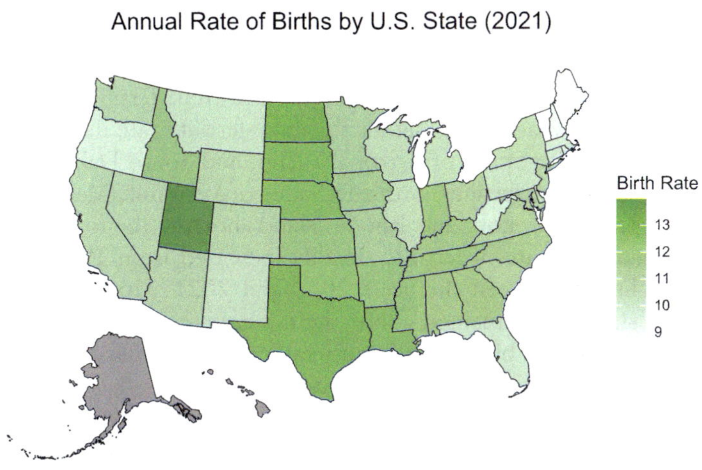

Fig. 6.7 Annual birth rate in 2021

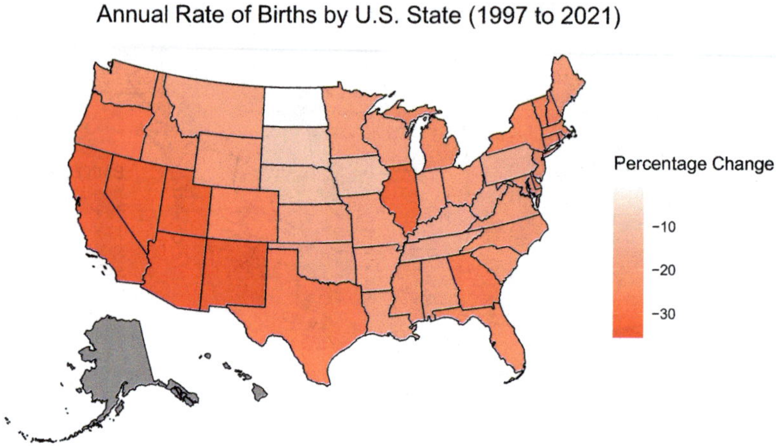

Fig. 6.8 Change in annual rate of birth

Figure 6.9 shows the state-level difference in the annual fertility rate, defined as the percentage of females aged 15 to 44 in the population, during the years 1997, 2010, and 2021. Utah has the highest fertility rate of all the states in the contiguous United States, with its fertility rate during 15 of the last 25 years being higher than any fertility rate in any year for any other state. The States of Arizona, Idaho, and South Dakota had the next highest rates of fertility, with the States of Vermont, Massachusetts, New Hampshire, Oregon, and Rhode Island showing the lowest fertility rates. Overall, the rate of fertility steadily increased from 1997 to 2010 before declining between the years 2010 and 2021. However, variance across the states has remained about the same, with fluctuations in both the annual number of births and fertility rates cutting across demographics throughout the contiguous United States.

Figures 6.10 and 6.11 show the state-level difference in annual fertility rates for the year 1997 versus 2021. The darker the color of green, the higher the rate of fertility. The lightest color of green indicates the lowest rate of fertility. Figure 6.12 illustrates the percentage change of the state-level difference for annual rate of fertility by US state for the years 1997–2021. The darker the color of red, the higher the percentage of change.

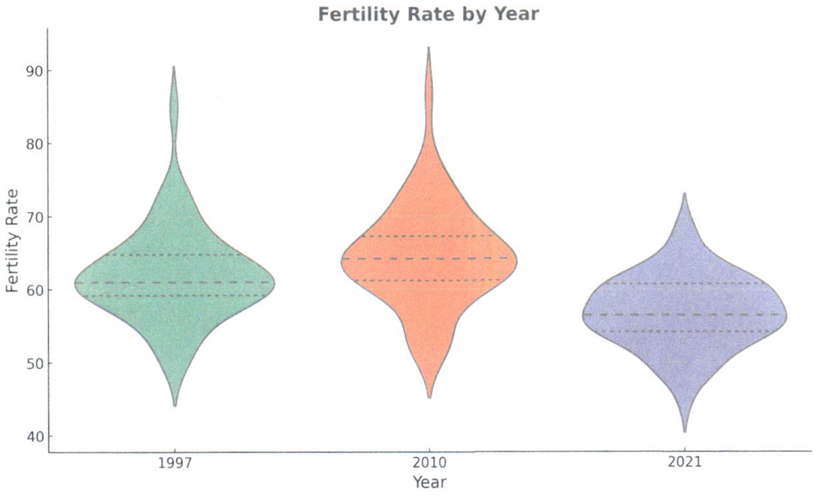

Fig. 6.9 Annual fertility rate

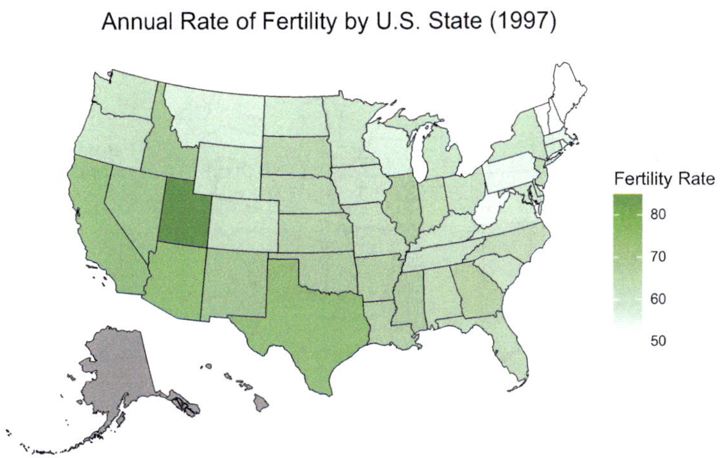

Fig. 6.10 Annual fertility rate in 1997

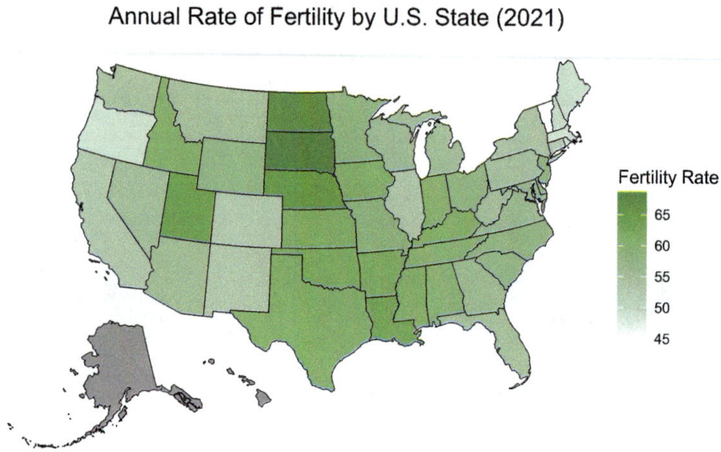

Fig. 6.11 Annual fertility rate in 2021

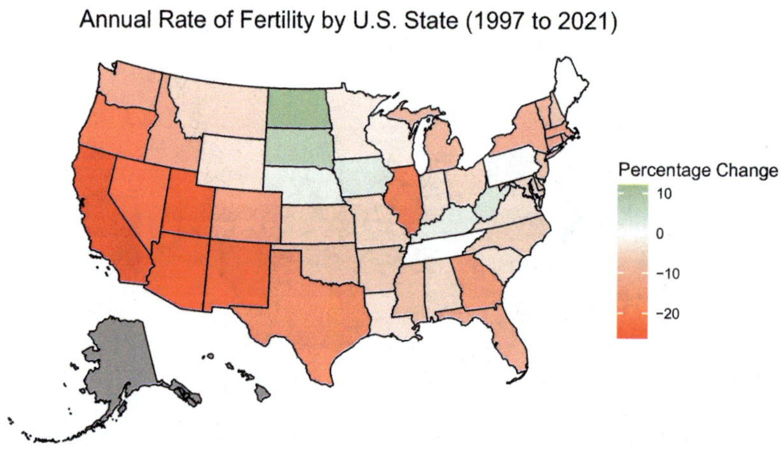

Fig. 6.12 Change in annual rate of fertility

In model 4, the combined expenditures for food, housing, and healthcare—referred to as the cost of living—have an inverse relationship with birth rates at the state level. Specifically, for each average increase of one unit of the rate of living costs, measured by consumer expenditures in millions of dollars divided by population in thousands of people, the average number of births each year decreases by 0.83 per 100,000 people. In addition, controlling for the rate of fertility, based on the percentage of females aged 15–44 in the population, birth rates and fertility rates are closely related. However, as the cost of living increases and fertility rates go up, birth rates are still expected to decline. For each average increase in the cost of living that is associated with a 0.83 decrease in the number of births per 100,000 people, there are approximately 22.8 more females aged 15–44 per 100,000 people in the population. Importantly, the decline in birth rates cannot be explained by changes in the fertility rate alone. These findings may help explain current research regarding dropping birth rates. Figure 6.13 illustrates the relationship between living costs, birth rate, and fertility rate at the state level across the years 1997–2009 and years 2010–2021.

In 2021, a group of researchers set out to explore the reasons behind the declining birth rate by analyzing several state-level differences, such as abortion, employment, healthcare, childcare, and debt, but were unable to identify any economic, political, or social factor, other than the Great Recession of 2008, to explain the drop in births between the years 2007 and 2020 (Kearney et al., 2022). Interestingly, the only factor the

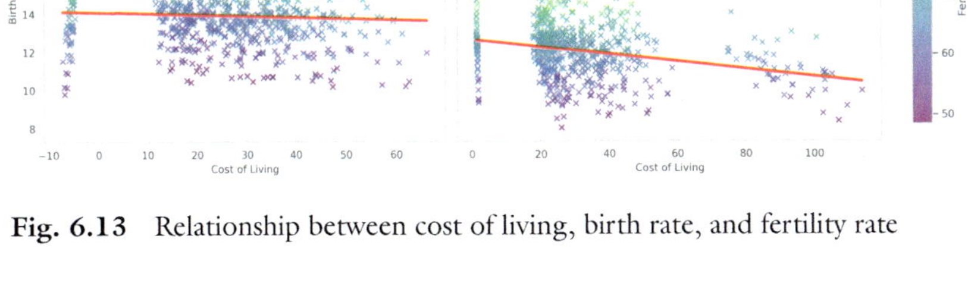

Fig. 6.13 Relationship between cost of living, birth rate, and fertility rate

researchers were able to identify was an economic factor, such as the cost of living, to explain the decline in birth rates for a few years, represented by a national economic recession. Researchers concluded that "shifting priorities," such as changes in fertility preferences, professional or personal aspirations, or parenting norms, may be to blame for the declining birth rate. Yet, the presence of environmental factors, such as rising costs of living due to extreme weather events and higher temperatures, was not considered.

In June 2025, the UN Population Fund (UNFPA) released its *State of the World Population* report, which focuses on the global decline in fertility between the years 1950 and 2021. Among the report's conclusions is a summary of findings from a recent survey outlining how many people do not expect to achieve their fertility desires due to economic pressures like the high cost of raising children and expensive housing. The report concludes that declining birth rates are not a result of younger generations rejecting parenthood. Rather, the report highlights how many people cannot have the number of children they desire due to factors like high living costs, such as housing, food and healthcare, and fears about the future, such as climate change (UNFPA, 2025).

Few studies explore or address how climate change positively correlates with an increased cost of living and, as a result, may influence the family planning decisions of individuals or couples, particularly decisions to have less children or no children simply because they cannot afford to feed, house, or care for an additional household member. Yet, how would family planning decisions change with increased knowledge that a family of five faces an increase in monthly expenditures of approximately $5000 due to increased food, housing, and healthcare costs? Further, would family planning decisions be impacted even more with the understanding that increased monthly expenditures drop to $4000 for a family of four? Or, $3000 for a family of three? What if every person entering adulthood and every couple embarking on the journey toward parenthood knew that increased monthly expenditures dropped to $2000 a month for a family of two and to $1000 for a family of one? How would an increased understanding of the relationship between our changing climate and the cost of living impact decisions around whether or not to have a child, along with how many children a household has the capacity to raise in a safe and healthy environment?

When increased costs of food, housing, and healthcare, driven by climate conditions, have an inverse relationship with birth rates, it can be inferred that climate change creates additional barriers to bodily autonomy. While we have already made the connection between climate change and environmental degradation, the findings of this study illustrate the significant impact a changing climate has on economic and societal conditions, including the limited ability for individuals to fulfill their fertility desires due to the increased costs associated with raising children. If rising prices prevent an individual (or couple) from freely choosing to have a child, then that individual does not have full bodily autonomy. Therefore, if climate change contributes to rising costs of living, which infringes on the ability to freely choose whether to have a child, climate change also infringes on the realization of reproductive justice.

The results of this study contribute to a growing body of evidence showing that climate change contributes to and adds complexities within societal problems. While climate change does not create food, housing, or healthcare insecurity, it may be contributing to it, along with strengthening vulnerabilities, adding challenges to remediation, and deepening ongoing crises. As such, climate change adds to the environmental and social issues that compound to create larger, more complex adverse impacts for communities at the intersection of climate change and social justice, particularly new and additional barriers to accessible healthcare, food security, and affordable housing, while also exacerbating vulnerabilities already present within our communities, further impeding and deteriorating individual bodily autonomy and the sustainability of healthy communities and vibrant economies. Therefore, climate change is detrimental to societal prosperity and in order to sustain healthy, vibrant societies, we must address and reverse our changing climate.

However, just as climate change does not impact us all equally, we are not all equally responsible for the greenhouse gas emissions fueling the warming of the planet. Deeply entangled within the conversation around the disproportionate impact of climate change on women's autonomy, the rising costs of food, housing, and healthcare, and barriers to reproductive justice, is the lavish lifestyle of the corporate elites, particularly powerful, high-paid white feminists. Modern corporate success models underpin and fuel many of the factors driving climate change through the emitting of greenhouse gases into the atmosphere that is contributing to global warming and the associated adverse impacts to soil moisture, surface temperatures, and extreme weather. This begs the question: if corporate expansion is driving climate change, are those who benefit from (and uphold) corporate success models responsible for the adverse impacts stemming from climate change? And if women, mostly low-income and women of color, are most harmed by climate change, what responsibility is assigned to the corporate white feminist who both benefits from the system driving climate change and claims to fight for gender equality?

References

Ip, G., & Adamy, J. (2024). Suddenly there aren't enough babies. The whole world is alarmed. *The Wall Street Journal.* Retrieved May 14, 2024, from https://www.wsj.com/world/birthrates-global-decline-cause-ddaf8be2

Kearney, M. S., et al. (2022). The puzzle of falling US birth rates since the Great Recession. *Journal of Economic Perspectives, 36*(1), 151–176.

United Nations Population Fund (UNFPA). (2025). *State of World Population 2025: The Real Fertility Crisis-The Pursuit of Reproductive Agency in a Changing World.* Stylus Publishing, LLC.

CHAPTER 7

The Rise of Corporate Colonialism and White Feminism

In 1963, Betty Friedan's iconic work *The Feminine Mystique* invaded the book clubs, nightstands, and minds of American housewives. In it, Friedan argued that, like a man, a woman's identity is cultivated through a sense of achievement, primarily through a career, and a woman's pursuit of professional and personal fulfillment should not be stunted by traditional gender roles. Within only a couple of decades from the book's release, the white affluent housewife, who appropriately performed her domestic and reproductive duties and was hailed by the West as the envy of women throughout the world, would now be measured against a new standard, one that would eventually be projected on all women around the world: an empowered woman gracefully carries her reproductive and domestic labor while also attaining the same professional success and status of her husband.

More than 50 years later, in 2015, Sheryl Sandberg would publish her book *Lean In* and quickly develop a cult following of empowered women. Shortly after, in 2019, Melinda Gates would release *Moment of Lift*, an ode to women empowerment through birth control and gainful employment for all women. Sandberg and Gates have long been lauded as advocates for gender equality and women's rights. However, these works, while widely popular, are examples of how women's liberation continues to be defined and measured by attaining the status and privileges afforded to affluent white men.

S. Haley Knowles, *Diminished Prosperity*,
https://doi.org/10.1007/978-3-032-07068-5_7

Critiques of the historical white woman-led charge for women's rights are not new. The literature is rich with testimonials and examples of how the women's rights movement focuses on the experiences of affluent white women and ignores the needs of nonwhite women. Critical race feminists, most importantly Loretta Ross, Dorothy Roberts, and bell hooks, have long called for white women to acknowledge and condemn their role within the historical and ongoing silencing of nonwhite women. What is less understood, however, is how the movement not only ignores the needs of nonwhite women but also contributes to and sustains their oppression.

As discussed in Chaps. 1 and 3, the reproductive rights movement grounds its fight in access to contraceptives and abortion mostly due to a cis-gendered white woman's ability to compartmentalize her healthcare and what she deems as gender discrimination based on her lived experiences. Unlike women of color and historically marginalized communities, white women rarely experience reproductive oppression based on other parts of their identity, such as race or immigration status. This is why the white women-led reproductive rights movement has such a narrow focus and does not provide the roadmap to reproductive freedom for all, as outlined by the reproductive justice framework. Further, because the systems of oppression that largely impact nonwhite women can be conveniently ignored by white women, white women are often implicated, albeit unintentionally, in the sustainment and advancement of these systems, such as corporate colonialism. This chapter examines modern corporate expansion as a continuation of settler colonialism and the role high-paid corporate women play in the modern oppression of marginalized women. The impact of corporate colonialism on the bodily autonomy of marginalized groups of women is discussed through the role of corporate expansion, white feminism, consumerism, and population control.

The Continuation of Colonialism by Corporate Expansion

In 1997, Edward Goldsmith published a poignant essay on the harms associated with Western development. Goldsmith argues that "development is just a new word for what Marxists called imperialism and what we can loosely refer to as colonialism" (Goldsmith, 1997, p. 69). While formal colonialism ended, colonizers did not end the practice of colonizing because they were no longer interested in the economic advantages it provided. Instead, colonial powers saw more effective and politically acceptable methods. Accordingly, international development efforts poised as philanthropic campaigns are, instead, a way to expand the West's market for goods and services through the acquisition of cheap labor and raw materials. A continuity exists "between the colonial era and the era of development, both in the methods used to achieve their common goal and in the social and ecological consequences of applying them" (Goldsmith, 1997). Therefore, it is no surprise that during the 1920s, when Henry Ford needed rubber for the tires of his automobiles, he looked at the Amazon rainforest as an opportunity for Western development.

In the late 1920s, the Ford Motor Company negotiated an agreement with the Brazilian government, resulting in 2.5 million acres of land in the Amazon basin being allocated for the establishment of Fordlândia, an industrial town with the goal of cultivating rubber and exporting it to the United States. The failure of Fordlândia is credited to numerous factors, including a shortage of labor due to the spread of diseases from Americans to local Indigenous workers and others refusing to work due to work conditions. Local workers were provided with American food, such as canned goods, and forced to reside in American-style housing, both of which were unfamiliar and uncomfortable to them. Further, workers were required to wear American identification badges and maintain American work hours during the height of the sun's heat, a departure from their customary work structure. Less than two years after the town's establishment, local workers began to revolt, cutting telegraph wires, chasing managers off properties, and kidnapping the town's cook. Ultimately, the town of Fordlândia was abandoned, but not before launching a predatory path for control, destruction, and exploitation in the Amazon. As a form of corporate colonialism, the legacy of Henry Ford contributed to the rise of neocolonial economics and ecological collapse in the Amazon basin that can be traced back to the community of Fordlândia (Barkemeyer & Figge, 2012).

Ford Motor Company was not the only proponent of corporate expansion and Western development during the 1920s. Around the same time that Ford was setting up his new town in Brazil, the United Fruit Company was leveraging the ideology of white supremacy to take domination over Indigenous land, resources, and culture and racially divide the labor of banana harvesting throughout Latin America. Like Fordlândia, United Fruit established American-style towns and housing for local workers, which were managed by American white men who exploited local workers, reinforced the racial hierarchy, and reproduced gender norms (Martin,

2018). While the impacts of colonialism on tribal nations of the Amazon differ from those of North America, the crimes of American companies leveraging the ideology of white supremacy to take domination over Indigenous land, resources, and culture are not unique to any one country or continent. This form of imperialism intersects with other forms of oppression to create crimes of the powerful and continue the legacy of colonialism. Corporate colonialism is the continuation of settler colonialism through the legitimized policies and practices of powerful nations taking and maintaining control over less powerful countries by establishing settlements and exploiting natural resources (Comack, 2018).

Today, corporate colonialism is fueled by capitalist culture and consumers' demand for their beloved avocados, almond milk, and smartphones, often sourced from developing countries. Capitalist culture is the set of social practices, behaviors, and values in a society that is built around an economic system that rewards increased production and consumption, the expansion of industrial activity, private enterprise, and a social hierarchy based on income (Comack, 2018). Consumerism is both an effect and a cause of capitalism. Without consumerism, capitalism would not exist, but without capitalism, consumerism would not be powerful. Consumer behavior is what drives an economy, particularly a capitalist economy. Within a capitalist society is a social hierarchy based on income and unjust class systems built upon white supremacy, settler colonialism, and the exploitation of land, natural resources, and low-income laborers.

The Role of White Feminism and Consumerism

An important component of modern corporate colonialism is white feminism, a racist ideology that began in the early 1900s with the rise of suffragettes and "Votes for Women." In 2021, theorist Koa Beck released her groundbreaking white feminism theory, which explores the role of not only race and class in feminism, but also the role of consumerism. According to Beck and rooted in capitalism and the attainment of the same privileges known to (white) men, white feminism lauded suffragettes as affluent white women who advocated for gender equality and women's right to vote through direct action and civil disobedience. Suffragettes yearned to be seen as equal to (white) men and believed that the best path to equality for (white) women was one that pacified their husbands and upheld Christian values and traditional family roles. Immigrants and other women who were poor, uneducated, fat, or those who identified as Black,

Indigenous, trans, queer, or any identity other than the patriarchal image of a "good white woman" were seen as a threat to the suffragette's agenda (Beck, 2021).

As a result, suffragettes created an image of who the ideal suffragette was and excluded anyone who did not meet their standard. A suffragette was aristocratic, fair-skinned, straight, thin, and well-educated. A suffragette strived to be a good wife, mother, and Christian woman. She supported business, dressed well, and enjoyed spending money (Beck, 2021). Therefore, the suffragette movement was not only comprised of an elite group of upper-class, white women, it was also a brand. In many ways, the white feminist identity conflates with the ability to wear luxury clothes and shop at exclusive stores (Beck, 2021). The ideology preaches the importance of individual success and conceives of equality as something women can achieve primarily through professional endeavors. At its core, white feminism is fundamentally exclusionary.

One of the main goals of white feminism is becoming an empowered elite woman with purchasing power, which often leads to the exploitation of nonwhite women (Beck, 2021). Accordingly, over a century later, one of white feminism's most loyal allies is consumerism. Today, white feminism is often promoted through a white corporate success model that has picked up momentum under the hashtag #girlboss. First introduced by Sophia Amoruso in 2015 in her book *#Girlboss*, the term represents the professional, successful, feminine woman empowered by her high-earning, independent status. These savvy white women executives leverage feminism as a marketing strategy, while also striving to not question or change systems of power, such as white supremacy, patriarchy, or settler colonialism, but instead to reach the highest level of success within existing structures, ignoring the harms these structures inflict upon other (nonwhite) women.

Take, for example, Sandberg's explanation in *Lean In* of why there are not more (white) women leaders in corporate America. She argues that boardrooms are dominated by men not due to systematic inequality or structural barriers, but because of the individual choices and actions of women. In other words, to solve gender inequality, women must change. Sandberg leverages her experience as a high-paid corporate executive to advise her readers that to change the power dynamic in corporate America, we must "lean in" to male-dominated executive spaces and marry a husband who helps with household chores and child-rearing (Daniels, 2015). *Lean In* reminds us of white women's aspiration to achieve the status and privilege of white men. The idea that working harder and marrying well is how we solve gender oppression not only reeks of white feminism but is also reminiscent of the suffragette agenda. By failing to acknowledge her privilege as an affluent white woman, Sandberg offers a one-size-fits-all approach to feminism that only benefits her, those with her lived experiences, and those who profit from consumerism and corporate expansion.

Similarly, Eve Ensler's *One Billion Rising* campaign is built upon white feminist branding. Ensler, best known for her play, "The Vagina Monologues," launched her new campaign on February 14, 2013, or what she refers to as "V-Day" for victory, valentine, and vagina. To fight violence against women, Ensler's organization raised hundreds of thousands of dollars through donations, the selling of branded merchandise, and a call for women to publicly gather on "V-Day" to bring awareness to violence against women. However, Ensler failed to acknowledge the "Memorial March for Missing and Murdered Indigenous Women," an annual remembrance walk held each year on February 14. Since 1991, Indigenous women throughout North America have gathered to honor and raise awareness of the disproportional rate of missing or murdered Indigenous women (Daniels, 2015). Ensler's appropriation of Indigenous women's political acts is an example of how white feminists erase the experiences of nonwhite women in an effort to advance their own agenda, while also encouraging consumption for their own capitalist gains, a primary driver of ongoing corporate colonialism and Indigenous women's oppression. Therefore, this form of feminism is not about the liberation of all women. It is about capitalism and maintaining the comfort of the elite white woman.

Another lesson can be found in Gates' *Moment of Lift*, where access to birth control and increased employment for low-income women is presented as the solution to gender inequality. While access to contraceptives and opportunities for gainful employment are critical for healthy communities, the notion that gender oppression is solved by certain populations limiting reproduction and joining the labor force that sustains the markets of the West is reminiscent of social policy of the 1900s. Gates's approach to gender equality reminds us that birth control continues to be pushed on historically marginalized communities and purchasing power continues to be the measure of equality for white feminists. Disguised as philanthropic endeavors and community development, white feminists are selling corporate development, or the modern form of European violence-based colonialism that inflicted centuries of pain, suffering, and oppression, to those who are most harmed by corporate colonialism: nonwhite women.

White feminism is white supremacy in disguise that promotes the comfort and safety of affluent white women by only focusing on sexism and masking white women's participation in white supremacy. As a result, white feminism functions to keep nonwhite women from realizing egalitarianist gains unless those gains also benefit white women, such as taking on reproductive labor or domestic housework to allow white women to succeed professionally. White feminism's capitalist ideology measures success by the individual accumulation of wealth, which is sustained through rising consumption rates. Yet, without the intensification of corporate domination of land and resources and the resulting resource extraction, the rising consumption rates that exponential growth requires are not possible. Accordingly, to keep up with rising consumption rates, corporate expansion continues. Western consumer industries are ripe with injustices that not only exploit low-wage workers but also pillage the earth and deplete its natural resources.

Weaponizing Feminism in the Fight for Impoverished Population Control

Discussion around the connection between consumerism and environmental degradation is not new. As theorist Aja Barber (2021) points out, "the byproducts of our quest to consume are creating an environmental crisis." Decades of calls for better environmental conservation, investments in renewable energy, and increased upcycling have gained popularity in recent years. The environmental movement has achieved many successes, including international agreements to reach global neutrality by 2050. Climate campaigns, green slogans, and environmental policies are well-received around the world and have even become trendy throughout Western society. Yet, acknowledgment of and accountability for who is most responsible for the environmental crisis seems to be less popular.

Studies show that longer work hours, higher income, and more wealth generally lead to higher energy consumption and carbon emissions. For example, the United States houses less than 5% of the world's population but is responsible for most global emissions released over the last century, and when organized by income, the top 10% of the world's population is responsible for more than 35% of annual global emissions (Briscoe et al., 2021). In addition, research demonstrates that households with more work hours, which equates to higher wages, have significantly higher rates of emissions due to higher rates of consumption (Fremstad et al., 2019). Further, as consumption levels increase and corporate expansion persists, climate change worsens. Therefore, the branded corporate white feminist appears to be a driving force of climate change. Yet, if successful white women leading the charge for gender equality are implicated in ongoing environmental degradation, what does that mean for feminism? In order to answer this question, we must first understand who is most harmed by climate change.

As previously discussed in Chap. 1 a rich body of literature demonstrates that climate change impacts are disproportionately felt by historically marginalized communities. The impact of emissions on a person's well-being is influenced not only by income, but also by race and gender, with historically marginalized communities being impacted the most and white women being impacted significantly less than nonwhite women (Briscoe et al., 2021). In addition, the impacts of climate change are experienced through the social structures present within a community. When extreme weather is combined with ongoing colonialism, it exacerbates the vulnerabilities already plaguing historically marginalized communities, such as a lack of affordable housing or food insecurity. Further, research shows that nonwhite women are at the highest risk of dying from climate-related or linked events, illustrating yet another example of how white feminism and its promotion of consumerism through corporate branding contribute to the oppression of nonwhite women (McDermott-Levy et al., 2021). Ironically, Sandberg's advice to work harder (or longer) as a solution to gender inequality and Ensler's obsession with feminist branding (or buying more stuff) to draw attention to women's oppression appear to do little more than maintain white women's comfort and drive climate change, which ultimately leads to further oppression of nonwhite women.

While white feminists' contributions to climate change are easily recognized, the harm to nonwhite women through white feminist branding and consumerism does not end with increased greenhouse emissions. In recent years, advocacy for increasing access to contraceptives to combat climate change, particularly in developing countries, has picked up popularity. Beginning in 2007 and continuing as recently as 2014, IPCC reports on human population growth as a primary cause of climate change is a catalyst for the promotion of population control policy as not only an approach for economic progress but also for environmental conservation (Cafaro, 2012). Further, the IPCC's conclusion that the majority of population growth is taking place in underdeveloped countries encourages population control policy to focus on low-income, communities of color in underdeveloped countries, or in other words, the global South.

At the core of population control policy is the goal to lower lifestyle carbon emissions, which, in theory, helps to combat climate change. However, as history shows, population control policy targets mostly low-income, nonwhite communities, those least responsible for climate change, based on both consumption and lifestyle emission rates, and those least likely to benefit from the industrial activity driving climate change, based on both income levels and buying power. Further, based on consumption levels, nonwhite women are least responsible for climate change while also being least likely to benefit from the economic processes driving climate change, based on income, wealth, and occupation, and least likely to participate in policymaking around climate solutions, based on positions of power (Briscoe et al., 2021). As a result, those most responsible for the lifestyle carbon emissions that are driving climate change—mostly affluent, white Western households—are not targeted by population control policy. Therefore, low-income, nonwhite communities continue to be scapegoated for global problems created largely by Western society, which allows responsibility to be deflected from the cushy lifestyles of the white elite.

A quick read of her book will find that Gates's gender equality campaigns target low-income, nonwhite communities, claiming to empower women by providing them the ability to control their fertility and opening up opportunities for education and employment, both of which are strong measures of success within the white feminist model. Ironically, Gates's approach to liberation for nonwhite women is to join the Western capitalist institutions responsible for much of the oppression nonwhite women experience, which begs the question: does population control policy really empower nonwhite women? If consumerism and corporate expansion drive climate change, and nonwhite women are most harmed by a changing climate, women empowerment campaigns that are rooted in girlbossing and the attainment of corporate success and purchasing power, really only benefit affluent white women.

As inconvenient a truth as it is, knowledge of the causes and drivers of climate change, along with an understanding of its systemic distribution of power and consequence, demands that corporate white feminists rethink their approach to feminism. While mainstream feminism and the reproductive rights movement have achieved many victories for gender equality—from voting rights to equal pay to improved access to birth control and abortion care, it is clear that some women have benefited from these victories more than others. Further, critiques of the white women-led reproductive rights movement do not negate the success, the progress, nor the good the movement has done for society. It is possible to celebrate the many freedoms that the reproductive rights movement has rightfully fought and won for women while also condemning the harms the movement has caused to many nonwhite women. Both of these truths can, and must, exist at the same time.

References

Barber, A. (2021). *Consumed: The need for collective change; colonialism, climate change & consumerism*. Hachette UK.

Barkemeyer, R., & Figge, F. (2012). Fordlândia: Corporate citizenship or corporate colonialism. *Corporate Social Responsibility and Environmental Management, 19*(2), 69–78.

Beck, K. (2021). *White feminism: From the suffragettes to influencers and who they leave behind*. Simon and Schuster.

Briscoe, M. D., et al. (2021). Intersectional indicators: A race and sex-specific analysis of the carbon intensity of well-being in the United States, 1998–2009. *Social Indicators Research, 155*(1), 97–116.

Cafaro, P. (2012). Climate ethics and population policy. *Wiley Interdisciplinary Reviews: Climate Change, 3*(1), 45–61.

Comack, E. (2018). Corporate colonialism and the "crimes of the powerful" committed against the Indigenous peoples of Canada. *Critical Criminology, 26*(4), 455–471.

Daniels, J. (2015, February 16). *The trouble with white feminism: Whiteness, digital feminism, and the intersectional internet. Digital Feminism and the Intersectional Internet.*

Fremstad, A., et al. (2019). Work hours and CO2 emissions: Evidence from US households. *Review of Political Economy, 31*(1), 42–59.

Goldsmith, E. (1997). Development as colonialism. *The Ecologist, 27*(2), 69–77.

Martin, J. W. (2018). *Banana cowboys: The United Fruit Company and the culture of corporate colonialism*. University of New Mexico Press.

McDermott-Levy, R., et al. (2021). Factors that influence climate change-related mortality in the United States: An integrative review. *International Journal of Environmental Research and Public Health, 18*(15), 8220.

CHAPTER 8

A Call for Indigenous Feminism

An important foundation of the white feminist model is the belief that all women derive the bulk of their oppression from patriarchy. Yet, patriarchy was a product of European society and foreign to most precolonial tribal nations. Many precolonial Indigenous women held political, social, and economic power that contradicted the European notion that women were second-class citizens, inferior to men, and the property of their husbands. While Indigenous women need liberation from patriarchy due to Christian capitalism brought by colonial agents, patriarchy is not the sole or even primary source of their oppression. However, with the white feminist agenda rooted in the power positions of affluent white settler women, sexual liberation, not decolonization, is the primary goal of white feminism. As such, there is much to be learned about reproductive justice from Indigenous feminists and the matriarchal traditions of tribal nations. This chapter explores the historical origins of Indigenous feminism and advocates for a reconstruction of mainstream feminism toward one that reimagines what prosperity means and takes a stand against white feminism's marriage to white supremacy, capitalism, and colonialism. The benefits derived from Indigenous lifeways are discussed through decolonialism, a reciprocal relationship to land, and a culture of giving.

S. Haley Knowles, *Diminished Prosperity*,
https://doi.org/10.1007/978-3-032-07068-5_8

The Role of Decolonization in Feminism

Although the call for decolonizing feminism is often muffled by girlboss noise, it is not an entirely new siren. Critical race feminists and Indigenous scholars and philosophers have been sounding the alarm for years. For example, in 2017, philosopher Margaret McLaren published the anthology *Decolonizing Feminism: Transnational Feminism and Globalization* to sum up this important body of work. The book rightfully acknowledges the groundbreaking work of many pioneers in this space, including Anibal Quijano and Chandra Talpade Mohanty, among others. Most notably is Quijano (2000), who, almost 25 years ago, identified the structural relationship between race and labor, mediated by capitalism and globalism, that gives rise to the exploitation of workers. Building upon this idea, Mohanty (2003) argues that a healthy economy, racial equality, and fair distribution of wealth are crucial elements in decolonizing feminism. Both of these works, along with others, provide the primer for why and how the white feminist model fuels consumerism and corporate expansion, and ultimately climate change and the resulting oppression of marginalized communities.

In 2020, McLaren joined forces with Celia T. Bardwell-Jones to edit a special issue of *Hypatia*, a similar anthropology that compiled various feminist theories addressing race, class, colonialism, and Indigenous lifeways "to cultivate and encourage theorizing about Indigenous philosophies and decolonizing methodologies" (2020). While the series has several notable contributions that highlight the many reasons Indigenous scholars reject mainstream feminism as a mode for advancing prosperity for First Nations, the most valuable contribution within this work may be the editors' acknowledgment in the introduction that feminist philosophy is mostly written by and for white women. The editors call out their privilege as white settler women and remind readers of the potential harm, although almost always well-intentioned, of settlers attempting to address the harm of a system they benefit from and uphold, and instead perpetuating further instances of the white feminist ideology (Bardwell-Jones & McLaren, 2020).

Accordingly, while my research aims to examine the role and participation of white women (including my own) in the white settler state and how we often benefit from the resulting ongoing oppression of nonwhite women, I recognize and acknowledge my limitations in identifying, addressing, and remediating the harms of the white settler state; for I, too, only became aware of the marriage between feminism and white supremacy, colonialism, and capitalism after reading the works of critical race feminists and Indigenous scholars, such as Elizabeth Cook-Lynn, a member of the Dakota and Crow Creek Sioux Tribes, who wrote *Who Stole Native American Studies?* to address colonialism as an ongoing state, not an isolated event of the past; or Myrna Kay Cunningham Kain, a member of the Miskito Tribe in Nicaragua, whose *Indigenous Women's Visions of an Inclusive Feminism* reminds us of the transformation needed for the white women-led feminist movement to be the liberation movement it claims to be. Others include Leanne Betasamosake Simpson, a Mississauga Nishnaabeg, and Leslie Marmon Silko, a woman of Laguna Pueblo descent, both groundbreaking Indigenous feminists whose contributions include calling for place-based indigenous alternatives to the settler colonial state (*As We Have Always Done,* 2017) and addressing the harms of white cultural imperialism within modern feminist ideals (*Yellow Woman,* 2013).

Indeed, there is much to be learned from Indigenous scholars about the role of decolonization and decentering the white elite in order for true feminism to be realized. Central to Indigenous lifeways are the ideas of decolonization, sovereignty, and reciprocal relationship to land. Decolonization is essential to Indigenous feminism, as much of the historical oppression Native women experience is the result of settler colonialism, which stripped tribal nations of their sovereignty, land, and cultural identity. Indigenous feminism grounds the concepts of autonomy and agency in the reciprocal relationship tribal nations have with the land, from which they derive their belonging, community, responsibility, and purpose. Thus, an important difference between white feminism and Indigenous feminism is how prosperity is defined. While white feminism measures success through the attainment of property, a symbol of status and power, Indigenous feminism condemns dominion and exploitation of land. Importantly, prior to the arrival of settlers, First Nations viewed land as communal, with an obligation to protect and share its benefits, which directly contradicted the white settler's concept of land as individual property. The very concept of land as a resource to own is a product of colonialism (Goldsmith, 1997).

The idea of land as property did not exist in the Americas during precolonial times. Rather, "in the process of settler colonialism, land is remade into property and human relationships to land are restricted to the relationship of the owner to his property" (Tuck & Yang, 2021, p. 5).

Accordingly, Indigenous Nations do not view land as a resource for development or property, nor the attainment of material things as a measurement of their success. This is why in 1979, when a federal court ruled that the United States government illegally seized the Black Hills from the Sioux Nation in 1862 after discovering gold in the region and awarded the tribe $17.5 million in damages, the Sioux refused to accept the money because the return of the sacred land was seen as much more valuable. Understanding how the Sioux measure prosperity also helps to explain why they continue to fight for the return of the Black Hills, with the award money sitting untouched by the tribe in an account managed by the Bureau of Indian Affairs, accruing compounded interest and totaling approximately $1.3 billion in current dollars (Lazarus, 1999).

Instead, many tribal nations define prosperity through the sustainment of healthy communities and land that is well cared for. Severt Young Bear, a member of the Lakota tribe, explains through his research how the Lakota Giveaway Ceremony is rooted in his tribal identity and relationship to the land. In Lakota tradition, when a member of the tribe reaches a certain level of success, they are honored in a ceremony where they give back to their community. The ceremony is based on a few core beliefs: possessions are not what life is about, what you give away comes back to you, and the success of a person's life is based on what you are able to give back to your family and community. The purpose of the Giveaway Ceremony is to share blessings with one's community and to release people's spirits from the attachment to material possessions (Bear, 1996). Many Indigenous tribes have a tradition similar to a giveaway in some form, often called a potlatch. The ceremony demonstrates how tribal nations define success not on degrees or corporate titles but on what they are able to give back to their community, which represents many First Nations' belief that taking care of each other and the land is the greatest demonstration of a prosperous life.

Tribal ceremonies and rituals, often at the core of First Nations' identities, are contingent not only on their community but also on the land where they are enacted. Rituals function as a reproduction of the culture, beliefs, and identity of Indigenous Peoples and the ability to pass rituals and ceremonies on to their children. Further, as scholar Stephanie Lumsden, a member of the Hupa Tribe, argues, "Indian Peoples' struggle against environmental racism and degradation of their lands is one way in which they assert tribal sovereignty and power over their fertility" (2016, p. 36). Therefore, an assault on tribal land is an assault on the reproductive ability of Indigenous Peoples. Sovereignty cannot exist without tribal nations maintaining their identities, which is tied to their relationship with the land. Tribal sovereignty cannot occur without tribal nations first reclaiming their right to self-govern and their right to the land, making decolonization essential to the liberation of not only Indigenous women but all Indigenous Peoples. As Dakota scholar Kim Tallbear (2012), a citizen of the Sisseton Wahpeton Oyate, writes, “the Dakota War of 1862 may be [the Dakota’s] most important origin story,” crediting the war with a “bloody re-mapping of Dakota life” and changing forever the identity of the Dakota Peoples.

Indigenous scholars teach us that unique to tribal nations are “those who have creation stories, not colonization stories, about how they came to be in a particular place” (Tuck & Yang, 2021, p. 6). Anthropologist Audra Simpson, a citizen of the Kahnawà:ke Mohawk Nation, argues that a settler’s possession of land is not possible without sustaining the dispossession of another. Simpson explains that because the identities and cultures of Indigenous Peoples are inextricably linked to the lands upon which they live and the natural resources upon which they depend, the continued possession of settler territory requires ongoing erasure of the tribal nations that belong to the land (Simpson, 2011). In the model of settler colonialism, the settler does not forfeit the land they or their ancestors took control of, resulting in white settler states depriving tribal nations of reclaiming their creation stories and realizing their true identity. Yet, decolonization requires not only the return of land to tribal nations but also the recognition of the sacred relationship between Indigenous Peoples and their land, viewed not as property but as an explanation of who they are. Therefore, with tribal nations still separated from their land through federal policy and corporate expansion, settler colonialism is ongoing, and Indigenous Peoples continue to be erased by the modern institutions formed out of the settler state (Simpson, 2016).

A Reciprocal Relationship to Land

Scholar Glen Coulthard, a member of the Yellowknives Dene First Nation, and his theory of grounded normativity call for a return to land-based Indigenous practices based upon the idea that the identity of humanity is inherently connected to nature. The Dene Peoples of the Northwest Territories in Canada are aware of both the inherent value of the coexistence of humans and nature and how important land is as a teacher.

Grounded normativity recognizes a reciprocal relationship between humans and nature and views the land as a model for balanced, harmonious living without domination or exploitation. The Dene Peoples value their land not only for the resources that their material survival depends upon but also for its significance to their identity and to their relationships with both other people and nature. Further, many tribal nations, including the Dene, believe that nonhumans embody agency in the same way that humans do, which requires humans to provide land, animals, rivers, and trees with the same respect we provide to other people. Grounded normativity believes that if we honor our obligation to nature, then nature will reciprocate and honor its obligations to us, which will provide for our sustained survival (Coulthard, 2014).

Because reciprocity to the land is central to cultural reproduction for First Nations, both the removal of a tribe from their ancestral lands and the destruction of tribal land is an attack on the tribe's identity. As Indigenous culture erodes and native lifeways are replaced by the white settler state, the children of First Nations are removed from their homelands to be taught by white settlers, resulting in centuries of history, culture, and customs, those that can only be passed down through land-based practices and rituals, being lost. This is why, when more than 60 members of the Osage Nation were murdered by white settlers who sought to inherit their land rights and, in turn, their wealth, often by marrying Osage women and then slowly poisoning them, the Osage mourned both for the loss of their loved ones and because their children would now be taught by white people. The Osage murders, which took place in Oklahoma during the first quarter of the twentieth century, are tragic not only due to the great loss of Osage life but also due to the inability of those killed to tell their children the stories of their elders and to pass down to their children the sacred ways of their people and their ancestral land, central to the identity of the Osage and passed down generation by generation in order to preserve the culture and ways of the Osage people (Grann, 2017).

Indigenous feminism requires not only the protection of bodily autonomy and freedom from violence for Indigenous women but also the protection of indigenous land. Yet, the white settler state is in direct contradiction with this ideology. As Kim Tallbear (2016) reminds us, "The white supremacist clearing of the land for white settlement, replacement, and control of the land is an ongoing project." Take, for example, in 2015, the Standing Rock Sioux Tribe, a sovereign nation of the Lakota, the people of Oceti Sakowin, contested the construction of the Dakota

Access Pipeline (DAPL), a new pipeline that threatened not only the tribe's water source but also went directly through sacred Lakota lands within unceded territory of the original Great Sioux Reservation. The West's quest for prosperity (oil) not only pillaged land held sacred to the Standing Rock Sioux Tribe, but it also threatened tribal identity (land). DAPL eventually was constructed and became operable, but not before a prolonged resistance of the Great Sioux Nation that drew violent-based control and removal tactics by the state in the name of legitimized economic progress, reminiscent of the violence-based practices of early settler colonizers and Indian agents (Gilio-Whitaker, 2019).

If Indigenous women's humanity is tied to land, the parallels between forced sterilization and assault on Native women's bodies and the raping of the earth by colonizers cannot be ignored. This relationship is further examined by Sarah Deer (2015), a member of the Mvskoke Nation, whose research explores the precolonial context of sexual assault. Prior to the arrival of settlers, tribal nations condemned the objectification of women with adequate tribal laws, rituals, and customs that protected women from such violence but did not have a word for rape. In fact, one of the first Mvskoke laws addressing what is known as rape today was written down at the insistence of settlers to demonstrate the lawfulness of the Mvskoke tribe and consisted of 43 words, 6 of which vested authority to decide the punishment for the perpetrator with the victim (Deer, 2015). The precolonial constructs of sexuality and bodily autonomy awarded women full rights to her body and the use of it. This changed, however, when settler colonization brought with it Christian capitalism and many traditional tribal gender norms were lost to reeducation and assimilation (Deer, 2015). It can be argued that colonizers brought with them what is known today as "rape culture," rooted in patriarchy, control of women, and disrespect for nature, and that violence against women is not a result of untamed passion or hormones but, instead, is a learned behavior.

Indigenous scholars and critical race feminists teach us that feminism is inherently central to Indigenous lifeways. The protection of bodily autonomy, equal rights, and feminine liberation is deeply embedded in Indigenous cultures. In precolonial times, Native women had full control over their bodies, including their fertility. For example, dating back centuries, tribal nations had a number of effective contraceptive methods that were widely accessible to both women and men. Plants and herbs, including the Indian paintbrush plant used by the Navajo and stoneseed used by both the Navajo and the Shoshone tribe, were often made into a tea and enjoyed by women and men to decrease a woman's menstrual cycle, lower a man's sperm count, or prevent the fertilization of an egg. Further, if a woman became pregnant and elected not to give birth, other plants that would terminate the pregnancy were readily available to her (De Laszlo & Henshaw, 1954). Ironically, tribal nations, those targeted by colonial agents for coercive fertility management, were using birth control long before Western medicine had a name for it. Therefore, the idea that nonwhite women need white women to provide them with access to birth control in order to be empowered falls short.

Reclaiming Feminism

Contrary to the mainstream definition of feminism, which is driving today's white feminist branding of philanthropic outreach campaigns to low-income and nonwhite communities, Indigenous women do not need a white form of feminism to liberate them through the regulation of their fertility. Instead, their liberation resides within their own version of feminism, a decolonial feminism. As Choctaw scholar Devon Mihesuah (2000) teaches us, it is not possible for one feminist voice to speak for all women or even for all women within a race or class category, nor is it appropriate for white feminists to try. Yet, white feminists often struggle to understand how their one-size-fits-all approach to feminism harms nonwhite women because it assumes that Indigenous women have the same experiences and needs as all women. By doing so, white feminism erases the unique experiences of nonwhite women, particularly Indigenous women and those who derive much of their oppression from colonialism, not patriarchy. Instead, we must look to Indigenous feminist scholars and theorist Dian Million, a Tanana Athabascan, who understands how the personal narratives of Indigenous women shine a light on the trauma of violence-based colonialism in a way no one else can. Because Indigenous women feel their histories as well as think them, their testimonials are more than just words; they are political acts (Million, 2009).

White feminism does not speak for all women. The mere act of Indigenous women sharing their stories is a form of decolonialism that does more for Indigenous women's liberation than white feminists' birth control campaigns ever could. While reproductive health is an issue that all women are impacted by and concerned with, the sexual violence and reproductive injustice Indigenous women experience not only make their concerns distinct from those of non-Native women but also demand a solution that disrupts the white feminist corporate model. Therefore, liberation for all women cannot exist if the liberation (corporate expansion) of the elite white woman depends upon the destruction of the Indigenous woman's identity (land). Decolonization requires a complete transformation of white feminism that unsettles all to truly liberate all; one that recognizes the sovereignty of tribal nations, returns the land to its rightful owners, and acknowledges the unique experiences of Indigenous women and other nonwhite women.

Western prosperity has always been tied to the oppression of others. From European colonialism to eugenics and population control policy to corporate women's empowerment campaigns, many examples of this relationship exist throughout American history. By white feminists defining equality and success based on the historical approach to prosperity, they contribute to and sustain the ongoing oppression of nonwhite women through their proximity to the power systems that they depend upon for their own liberation. As a result, through corporate expansion and consumerism, white feminists create and maintain corporate colonialism, and ultimately the colonization of Indigenous women. Therefore, white feminism does not fight for the rights of all women but, instead, exists to protect the comfort of affluent white women. True feminism, or one that seeks to liberate all women, requires a decolonial feminist model. However, the decolonization of white feminism cannot occur in isolation or without the participation of all, for "settler colonialism and its decolonization implicates and unsettles everyone" (Tuck & Yang, 2021, p. 7).

A reconstruction of feminism is urgently needed, one that reimagines what prosperity means and takes a stand against white feminism's marriage to white supremacy, capitalism, and colonialism. We need a feminism that condemns white feminists' focus on sexual liberation as the main solution to women's empowerment, one that calls out the harms of the white savior complex, the white aid industrial complex, and the corporate girlboss in favor of a new form of prosperity and freedom. We need a feminism that goes beyond the liberation of white women from sexual and reproductive oppression and recognizes colonialism as the root cause of much suffering, one that demands that the movement be more than gender equality. We need a feminism to restore not only the economic and social dimensions of what women's liberation means but also the unapologetic, radical political dimension that decenters the individual success model of the white elite. The closest ideology to this goal may be that of Indigenous feminism or, as the Dene Peoples strive, one that "protects the intricately interconnected social totality of a distinct mode of life; a life on/with the land that stressed individual autonomy, collective responsibility, nonhierarchical authority, communal land tenure, and mutual aid, and which sustained us economically, spiritually, socially and politically" (Coulthard, 2014, p. 65).

References

Bardwell-Jones, C. T., & McLaren, M. A. (2020). Introduction to Indigenizing and decolonizing feminist philosophy. *Hypatia, 35*(1), 2–17.

Bear, S. Y. (1996). *Standing in the light: A Lakota way of seeing.* U of Nebraska Press.

Coulthard, G. S. (2014). *Red skin, white masks: Rejecting the colonial politics of recognition.*

De Laszlo, H., & Henshaw, P. S. (1954). Plant materials used by primitive peoples to affect fertility. *Science, 119*(3097), 626–631.

Deer, S. (2015). *The beginning and end of rape: Confronting sexual violence in Native America.* U of Minnesota Press.

Gilio-Whitaker, D. (2019). *As long as grass grows: The Indigenous fight for environmental justice, from colonization to Standing Rock.* Beacon Press.

Goldsmith, E. (1997). Development as colonialism. *The Ecologist, 27*(2), 69–77.

Grann, D. (2017). *Killers of the flower moon: The Osage murders and the birth of the FBI.* Vintage.

Lazarus, E. (1999). *Black hills white justice: The Sioux nation versus the United States, 1775 to the present.* U of Nebraska Press.

Lumsden, S. (2016). Reproductive justice, sovereignty, and incarceration: Prison abolition politics and California Indians. *American Indian Culture and Research Journal, 40*(1), 33–46.

Mihesuah, D. A. (2000). A few cautions at the millennium on the merging of feminist studies with American Indian Women's Studies. *Signs: Journal of Women in Culture and Society, 25*(4), 1247–1251.

Million, D. (2009). Felt theory: An Indigenous feminist approach to affect and history. *Wicazo Sa Review, 24*(2), 53–76.

Mohanty, C. T. (2003). "Under western eyes" revisited: Feminist solidarity through anticapitalist struggles. *Signs: Journal of Women in Culture and Society, 28*(2), 499–535.

Quijano, A. (2000). Coloniality of power and Eurocentrism in Latin America. *International Sociology, 15*(2), 215–232.

Silko, L. M. (2013). *Yellow woman and a beauty of the spirit.* Simon and Schuster.

Simpson, A. (2011). Settlement's secret. *Cultural Anthropology, 26*(2), 205–217.

Simpson, A. (2016). The state is a man: Theresa Spence, Loretta Saunders, and the gender of settler sovereignty. *Theory & Event, 19*(4).

Simpson, L. B. (2017). *As we have always done: Indigenous freedom through radical resistance*. U of Minnesota Press.
Tallbear, K. (2012). *The US-Dakota War of 1862: Ethnic strife or the political economy of war?* Oak Lake Writers Society.
TallBear, K. (2016). Annual meeting: The US-Dakota War and failed settler kinship. *Anthropology News, 57*(9), e92–e95.
Tuck, E., & Yang, K. W. (2021). Decolonization is not a metaphor. *Tabula Rasa*, (38), 61–111.

PART III

The Shortcomings of Classical Economics in a Modern Society: Evaluating Prosperity in the Era of Climate Change

CHAPTER 9

Traditional and Alternative Approaches to Prosperity

In September 2023, the USDT released a report highlighting how climate change is responsible for significant financial strain to American households by linking several financial impacts to climate-related risks and events, such as reduced income, increased destruction of property, increased transportation costs due to infrastructure collapse, added healthcare costs from climate-related medical needs, and higher expenditures on utilities from increased energy consumption. In addition, the report explains how climate change increases a household's need for savings and insurance products due to unpredictable climate events, along with the increased risk of gaps in insurance coverage as insurers increase household premiums, reduce coverage, or choose not to renew coverage due to climate risk (USDT, 2023). With increased knowledge of the relationship between global warming and costs of living, the implications of a changing climate to society continue to become clearer. Yet, few policymakers, community leaders, heads of households, or boards of directors make the connection between the climate crisis and social safety net programs, supplemental food vouchers, employee health benefits, or a volatile housing market.

This chapter explores how climate change directly impacts the prosperity of families, communities, and economies by impeding livelihoods through increased costs of living, exacerbating parenting costs, influencing family planning decisions, and preventing households, communities,

S. Haley Knowles, *Diminished Prosperity*,
https://doi.org/10.1007/978-3-032-07068-5_9

and economies from healthy, sustainable growth and resiliency. By examining how a modern society measures prosperity through a review of classical economic theory, historical policy approaches to promoting healthy families and communities, and modern corporate approaches to financial performance, the chapter leverages the theoretical framework developed through Chaps. 1–3, and the results of a new study that measures the relationship between outputs of climate change and components of reproductive justice, to call for a new prosperity model based on both financial and nonfinancial performance measures with equal consideration to society, the planet, and the economy.

The historical view of prosperity, or economic and societal progress, is best discussed through the views of the three Fathers of Economics: Adam Smith, who advocated for free markets; Karl Marx, who advanced a radical socialist model; and John Maynard Keynes, who believed in big government and the welfare state. However different in their approach, all three criticized mercantilism, the nationalist economic theory of their time that promoted one-sided trade to maximize resource accumulation within a country, as a limit to human prosperity (Chijioke et al., 2021).

The Classical Approach to Prosperity

Adam Smith, considered by many to be the Father of Microeconomics and the founder of capitalism, believed that prosperity is comprised of three components: human labor, contractual exchange, and preserved self-interest, with an emphasis on the division of labor. In his classic work, *An inquiry into the nature and causes of the wealth of nations* (1776), Smith argues that "labor…is alone the ultimate and real standard by which the value of all commodities can at all times and places be estimated and compared" and establishes the groundwork for free markets in economics. Smith promoted an economics of scarcity, advocated against taxes, and believed rational self-interest and competition lead to economic prosperity or, in other words, that by doing what is best for oneself, an individual does what is best for society (Smith, 1776).

In contrast, Karl Marx, an economist and philosopher sympathetic toward the proletariat, published his *Communist Manifesto* in 1848, which was followed by a three-volume *Das Kapital*, published over a period of several years from 1867 to 1894. In both, Marx criticizes capitalism as a mode of materialism and the idea that harmonious societies develop through class conflict as a system of oppression, exploitation, and, eventually, failure. Marx explains how capitalism creates tensions between classes, eventually leading to class consciousness and self-destruction. He believed capitalism would eventually be replaced by a society of individuals who, absent state, social class, hierarchy, or private ownership, would freely produce and reproduce the conditions of a society based on their individual needs and desires (Marx, 1907).

Somewhere in between the self-interests of economic internationalism and class-free Marxism resides Keynesian economics, known as a collection of ideas promoting governments with a responsibility to interfere with markets in times of crisis. During the Great Depression of the 1930s, John Maynard Keynes, known by many as the Father of Macroeconomics, published *The Means to Prosperity*, a series of essays challenging neoclassical economics by advocating for counter-cyclical public spending to address unemployment. Shortly after, in 1936, Keynes published *The General Theory of Employment, Interest and Money*, where he outlined how fiscal policies can be used to mitigate the adverse effects of economic recessions and depressions. Keynes believed that a sustainable, harmonious system of capitalism that promotes peace and prosperity was possible, but only if the system was managed with the freedom of trade and without pitting societies against each other (Keynes, 1936).

While wildly different in approach and theory, economists Smith, Marx, and Keynes were pioneers in the exploration, understanding, and promotion of prosperity as a measure of progress based on the establishment of societies and the development of economic institutions, activities, and participants. Today, modern economics and fiscal policy are rooted in ideals and principles established centuries ago, at a time when societies were shifting from the human economy and agrarian institutions of mercantilism to mechanical processes, factory systems, and industrial manufacturing, when economists had the luxury of measuring production and consumption without the consideration of climate change. However, following the end of World War II and the Industrial Revolution, countries, cities, and households across the globe strengthened their economies by increasing their reliance on fossil fuels, which begs the question: is the modern understanding of prosperity still relevant?

At the turn of the twentieth century, following the publication of Smith's and Marx's iconic works but preceding the advent of Keynesian policies, economist Simon Patten published *The Theory of Prosperity* (1902), highlighting his belief that a system of capitalism will always provide enough resources for everyone and that, while individuals may despair from unfulfilled pursuits for prosperity, anguish is due to the misuse of goods, not the lack of them. Patten builds upon his theory in *The New Basis for Civilization* (1968), where he advocates for a new civilization, one founded upon economic institutions that are efficient and productive, in place of a civilization developed from moral or ethical principles. Patten promoted an economics of abundance and believed that poverty was not sustained over the long term in a capitalist system as long as resources were utilized through efficient production and consumption.

Alternatively, social theorist Max Weber warns of a new civilization leading to the disempowerment of the modern self in *The Protestant Ethic and the Spirit of Capitalism* (1904). Weber introduces the idea of rationalization in the context of capitalism, believing that with the rise of modern capitalism, societies replace tradition and values as the main drivers of social behavior with concepts of reason and logic. And, while rationalization can bring enhanced individual freedom through increased knowledge from the ability to calculate and predict, Weber points out that agency is often crippled by capitalist systems through growing impersonality and enhanced control of humans, which leads to "sensualists without heart" and "specialists without spirit," the two sides of disempowerment. Most notably is Weber's prediction that the end stages of capitalism will bring about one of two scenarios: either re-enchantment with the birth of new deities or the rebirth of old ideals, or "mechanized petrification embellished with a sort of convulsive self-importance" (Weber, 1904, p. 183).

Alternative Approaches to Prosperity

Today, with rising concerns about the conflict between growing greenhouse gas emissions and human well-being, many believe American society resides within late-stage capitalism, where individual freedoms, once enhanced through mechanical and digital processes with expanded employment opportunities, lower consumer prices, and increased access to information, have deteriorated due to widening gaps between social classes and expansive corporate monopolies. For example, in 2016, ecological economist Tim Jackson outlined the conflicting relationship between human well-being and economic growth in *Prosperity Without Growth: Foundations for the Economy of Tomorrow*, first published in 2009 in a report to the United Kingdom Sustainable Development Commission. The book outlines evidence showing that, beyond a certain point, economic growth does not increase human well-being, while exploring the complex relationships between economic growth, environmental crises, and social recession. Jackson's proposed route toward a sustainable economy includes a restructuring of the definition of "prosperity" within the boundaries of current research that maps global warming to adverse impacts on humanity (Jackson, 2016).

Research demonstrating the link between climate change and human well-being is growing rapidly. For example, in 2018 alone, five studies were published illustrating the correlation between extreme weather events and the well-being of families, communities, and economies. One study concluded that as climate change increases, along with the associated increased incidence of climate disasters and damage to ecosystems, adverse impacts on economic systems, along with increased rates of food insecurity, infectious disease, and poverty, will also increase (Mal et al., 2018). Another study found extreme weather events not only adversely impact socioeconomic and environmental attributes, but also demographic factors, such as family planning and birth rates (Nandi et al., 2018). A third study found a correlation between climate disasters and income disparity between classes, with incomes for non-agriculture households above the poverty line being impacted the most by climate events (Keerthiratne & Tol, 2018). Two additional studies, which explored the relationship between climate change and progress toward the United Nations Sustainable Development Goals (SDGs), found global warming is a driver of increased occurrence of climate disasters and, in turn, decreased progress toward the SDGs, concluding the following: economic development is dependent on climate change mitigation, methods of economic planning should consider the economic and human costs of extreme weather events, and climate change mitigation requires investment in the development of sustainable infrastructure (Wu et al., 2018; Xie et al., 2018).

Building upon these studies, in 2019, a group of researchers selected seven Asian countries, namely China, Japan, India, Pakistan, Vietnam, Bangladesh, and the Philippines, to analyze the impact of climate disasters on indicators of human well-being and overall societal prosperity. The study reviewed data from the World Bank and the International Disaster Database, a database compiling information on the occurrence and impact of mass disasters, over a 13-year period between 2005 and 2017, and concluded that extreme weather events, defined as storms and associated floods, increase migration, consumer prices, and rates of poverty, which, in turn, decrease access to economic resources, measured by increased costs of healthcare and energy demand and decreased rates of financial development (Abbas Khan et al., 2019). Yet, as more evidence emerges demonstrating that historical approaches to prosperity, those that center on unchecked economic growth, fall short, many economists disagree on an effective, pragmatic path toward sustainable economic growth, or one that positively correlates with human and societal well-being.

In 2020, two theorists introduced a framework to address the dissatisfaction of individuals and communities left behind by globalization, technological advancement, and economic progress through the "decoupling" of economic prosperity and social prosperity. The framework consists of four indices: one representing social well-being, measured in terms of solidarity which covers the human need for social belonging; one representing individual well-being, measured in terms of agency, which covers the human need to control their own fate; one representing economic growth, measured in terms of material gain by Gross Domestic Product (GDP) per capita; and one representing ecosystem health, measured in terms of environmental sustainability by the Environmental Performance Index. Included with the framework is a theoretical analysis with findings that identify a correlation between both (1) solidarity and GDP and (2) agency and GDP, showing the correlation between both sets of indicators is in decline (Lima de Miranda & Snower, 2020).

A year later, in 2021, an inclusive prosperity framework, comprised of a taxonomy of public policies to address economic instability and inequity, was introduced. The framework divides policies into two dimensions based on the social class of the policy intervention's target beneficiary and the stage of policy production where the intervention takes place, organized by preproduction, production-stage, and postproduction policies. Importantly, the authors point out that traditional welfare states mostly rely on social policy in two stages to correct inequality and insecurity: preproduction policy, or interventions that shape market outcomes through endowments, such as education and vocational training, brought to market by individuals; and postproduction policy, or interventions that shape the distribution of capital through the transfer of wealth, such as progressive taxation and social insurance, after economic decisions have been made. Separately, production-stage policy, or interventions that shape the economic decisions of firms through increased competition and employment, is relied upon solely for increased productivity, innovation, and growth. The framework advocates for social and economic policy to merge; for social policy to more closely resemble industrial policy and economic policy to mirror labor market policy; and for policymakers to leverage not only preproduction and postproduction policies to address inequality, but also production-stage policy (Rodrik & Stantcheva, 2021).

Those reluctant to integrate social equity objectives into production-stage policy often fall into the self-described "socially progressive but fiscally conservative" camp; those who acknowledge the existence of, and the harm caused by, systemic oppression of certain demographic groups, but balk at any perceived discomfort or inconvenience to them caused by proposed policy interventions aimed at calibrating the same oppressive institutions that give them a leg up. In fact, having an advantage seems to be expected, even normalized, as an indication of prosperity. For example, a recent review of modern economic theory explored whether, with the rise and promotion of capitalism globally, mercantilism has truly perished. Leveraging game theory, which, in 1953, was significantly expanded by mathematician John Nash's decision-making theorem to show a player's strategy is only optimal when considering both their optimal outcome and those of other players, researchers analyzed centuries of economic achievements through discovery, innovation, and technology to conclude that, absent the consideration of others, the associated economic strategies of colonialism, the industrial revolution, and globalization are actually evolvements of mercantilism (Chijioke et al., 2021). Consequently, in light of these conclusions, is it possible then to achieve societal prosperity within the constructs of modern economic theory?

For true prosperity to be realized, it must be understood as a component of freedom and, therefore, dependent upon the protection, promotion, and realization of individual autonomy. In 1785, Immanuel Kant published the *Groundwork of the Metaphysics of Morals*, where he argues that the ability of individuals to make rational decisions about their lives is the fundamental characteristic of human beings, highlighting the importance of respecting individual autonomy and treating each person as an end within themself, rather than a means to an end, as often promoted by mercantilism. This idea of autonomy was further explored by John Stuart Mill, who published *On Liberty* in 1859 to shine light on the importance of allowing individuals to make their own decisions about their own bodies and their own lives without interference from the state. Indeed, the path toward sustainable development is paved with autonomy, which is the foundation of individual satisfaction, communal well-being, and societal prosperity. Yet, how individual freedom and autonomy interact or correlate with familial prosperity is more complicated.

The Economy of the Family

To fully understand the health of a family, it is important to understand how society interprets familial prosperity. With a shift from agrarian societies to industrial ones, and the social transformation that accompanied it, the economic benefits of children mostly dissolved, while the cost of children, such as education expenses, increased steadily over time (Schoen et al., 1997). In 1960, Gary Becker, one of the first theorists to explore family economics, used an economic model to analyze fertility desires and concluded that children were like consumer goods, or, as income increases, so does the number of children within a family unit. When it was found that higher -income groups frequently have fewer children, Becker further explained that the demand for children depends not only on parental income but also on the cost of rearing children, with higher-income households desiring higher-quality children, which tend to be more expensive (Becker, 1960).

A couple of decades later, in 1981, Becker published *The Treatise of the Family,* in which he analyzes the institution of the family through the lens of household economics, explaining that the choice to have children is made by individuals maximizing their utility, or by comparing benefits and costs. Becker acknowledged that income and other limited resources restrict human activity but believed that the greatest barrier to humanity realizing its full potential was the constraint of time, highlighting his belief that the flow of goods and resources can satisfy everyone's needs, but not everyone will have their needs met due to limited time (Becker, 1992). Becker's limited conclusions on fertility desires, although important, are an example of how the role of bodily autonomy in determining the fulfillment of fertility desires is often ignored, forgotten, or misunderstood.

Understanding how climate change impacts the fulfillment of fertility desires helps to understand the impacts of climate change on the realization of prosperity for families, communities, and economies. Although the potential for interaction between population growth and economic growth has been on the agenda of economists for some time, the consideration of a changing climate within this relationship has not been considered. For example, in 1939, Alvin Hansen argued that, following the First World War, the low birth rates of the Great Depression created less incentive for economic investment, which resulted in decades of slow economic growth. Hansen's research built upon Adam Smith's theory that population growth increases productivity due to a larger population providing more opportunities for labor. However, decades after their theories were published, the onset of climate change and its associated impacts on financial stability, along with an increased collective awareness of the positive link between industrial activity and global warming, appear to have an opposite impact on population growth. Importantly, Hansen and Smith developed their hypotheses based on their understanding of economic and human activity without experiencing or understanding the consequences of a warming planet.

Unfortunately, the reality of climate change can no longer be ignored in economic policy. Climate change creates additional barriers to full bodily autonomy through increased costs of parenting caused by economic instability, which prevents both families from fulfilling their fertility desires and society from realizing prosperity. Historical economic models claim to provide the pathways toward prosperity, but most have neglected to consider what might happen if unforeseen obstacles obstruct the path. To be effective and relevant in modern times, economic theory must not only provide the formula for societal prosperity but must also protect the right of all people in a society to realize prosperity, or the right to have and raise healthy children in safe and healthy environments.

References

Abbas Khan, K., et al. (2019). Natural disasters and economic losses: Controlling external migration, energy and environmental resources, water demand, and financial development for global prosperity. *Environmental Science and Pollution Research, 26*, 14287–14299.

Becker, G. S. (1960). An economic analysis of fertility. In *Demographic and economic change in developed countries* (pp. 209–240). Columbia University Press.

Becker, G. S. (1992). Fertility and the economy. *Journal of Population Economics, 5*(3), 185–201.

Chijioke, O. C., et al. (2021). Mercantilism in perspective: A historic review. *Global Economy Journal, 21*(01), 2150001.

Jackson, T. (2016). *Prosperity without growth: Foundations for the economy of tomorrow*. Taylor & Francis.

Keerthiratne, S., & Tol, R. S. (2018). Impact of natural disasters on income inequality in Sri Lanka. *World Development, 105*, 217–230.

Keynes, J. M. (1936). *General theory. Reprint*. Harbinger.

Lima de Miranda, K., & Snower, D. J. (2020). Recoupling economic and social prosperity. *Global Perspectives, 1*(1).

Mal, S., et al. (2018). Introducing linkages between climate change, extreme events, and disaster risk reduction. In *Climate change, extreme events and disaster risk reduction: Towards sustainable development goals* (pp. 1–14).

Marx, K. (1907). *Capital: The process of circulation of capital*, ed. by Frederick Engels, tr. from the 2d German ed. by Ernest Untermann (Vol. 2). CH Kerr.

Nandi, A., Mazumdar, S., & Behrman, J. R. (2018). The effect of natural disaster on fertility, birth spacing, and child sex ratio: Evidence from a major earthquake in India. *Journal of Population Economics, 31*, 267–293.

Rodrik, D., & Stantcheva, S. (2021). *A policy matrix for inclusive prosperity (No. w28736)*. National Bureau of Economic Research.

Schoen, R., et al. (1997). Why do Americans want children? *Population and Development Review*, 333–358.

Smith, A. (1776). *An inquiry into the nature and causes of the wealth of nations: Volume One*. Printed for W. Strahan; and T. Cadell.

United States Department of the Treasury. (2023). *The impact of climate change on American household finances.*

Weber, M. (1904). *The protestant ethic and the spirit of capitalism.*

Wu, J., et al. (2018). Economic development and declining vulnerability to climate-related disasters in China. *Environmental Research Letters, 13*(3), 034013.

Xie, W., Rose, A., Li, S., He, J., Li, N., & Ali, T. (2018). Dynamic economic resilience and economic recovery from disasters: A quantitative assessment. *Risk Analysis, 38*(6), 1306–1318.

CHAPTER 10

The Right to the Realization of Prosperity

The full realization of prosperity requires first the realization of full bodily autonomy and the fulfillment of fertility desires in the face of growing climate anxiety and impacts. While obtaining full bodily autonomy looks the same for most, the attainment of fertility desires is quite personal. Societal, environmental, and economic factors may influence an individual's desire to have or not to have a child and how many children they wish to rear, but the right to realize one's fertility desires is second only to the right to one's own life. Yet, the journey to living out one's fertility desires has gotten more complicated over the last few decades with increasing societal pressures and economic externalities impacting one's ability to freely choose their reproductive destiny. Further, environmental threats are a growing concern among many of reproductive age, causing some to decide not to fulfill their fertility desires due to fears of climate change. This chapter explores the theories of structure and bodily autonomy to better understand the barriers to families, communities, and economies realizing prosperity. Discussed through the principles of full bodily autonomy, fertility desires, and climate anxiety, the impact of climate change on an individual's or a family's ability to realize prosperity is examined.

In 1992, William H. Sewell, Jr. introduced a new theory of structure through a more precise understanding of the relationship to duality, agency, and transformation. Through his work, Sewell sought to bridge the gap between anthropology's semiotics, or the study of language,

S. Haley Knowles, *Diminished Prosperity*,
https://doi.org/10.1007/978-3-032-07068-5_10

symbols, and communication within a society, and sociology's materialism, or the belief that all aspects of society can be explained through natural processes, to restore the concept of human agency to social actors (Sewell, 1992). Sewell's theory was largely in response to the work of Anthony Giddens, known for his structuration theory with an emphasis on the duality of culture, or the understanding that while people shape structure, structure determines what people do, and two important works, *The Constitution of Society* (1984) and *The Consequences of Modernity* (1990). While Giddens acknowledges the transformative nature of globalization on societies through interconnectedness, he highlights how global processes erode local cultures, along with exacerbating environmental degradation, economic inequalities, and political instability (Giddens, 1990).

Sewell expands on this idea by organizing structures based on two components, the power dimension, which encompasses resources, and the depth dimension, which comprises cultural rules and norms, and by arguing that both dimensions must "mutually imply and sustain each other over time" in order for a structure to exist (Sewell, 1992, p. 13). The introduction of resources into structure is important for a couple of reasons. First, Sewell argues that resource consequences are unpredictable in the advent of a cultural rule, and, second, that because resources mean different things to different people, leading to varying degrees of interaction, the impact to resources from changes in cultural norms is measured by both nature and interpretation. Sewell believed that structures overlap and that it is social actors, through their transformative capacity, that determine how rules are implemented and how situations are perceived. In the context of societal structure, Sewell's understanding of human agency is an "actor's capacity to reinterpret and mobilize an array of resources in terms of cultural schemas other than those that initially constituted in the array," which results in the empowerment of individuals to not only cooperate with others through structures but also to use structures against others or, in his words, "Agency arises from the actor's knowledge of schemas, which means the ability to apply them to new contexts" (Sewell, 1992, p. 20).

A decade later, in 2005, Sewell presents a controversial argument critiquing the classical approach to structure and culture by highlighting the importance of historical events in shaping structures over time, resulting in the measure of structural changes in the context of historical happenings being critical for understanding the relationship between society and institutions (Sewell & Sewell, 2005). Both Sewell's emphasis on the power of resources in social structures, along with the role of historical events and the social actors who shape them, and Giddens's explanation of the benefits and detriment of societal development can be viewed through Sen's and Nussbaum's capabilities approach. Here, poverty is considered a deprivation in a human's capability to realize a good life, while development is considered an expansion of humanity's capability to lead prosperous lives. Therefore, if a societal structure prevents a person from realizing their fertility desires, then does it also deprive them of their capability to lead a good life, measured by social structures in the form of access to resources, changes in cultural norms, and the social actors who distribute and interpret them?

Full Bodily Autonomy

For centuries, theorists have sought to understand the role of and impacts on autonomy. Kant, one of the first to highlight its importance, believed autonomy, which he defines as the moral right of an individual to make their own decisions about their own life, is one of two components needed for a meaningful life. The other is rationality. More than 100 years before Weber used rationalization to describe markets, Kant explains rationality as an individual's motivation to govern their own life (Kant, 1785). Building upon Kant's ideals, in 1881, Friedrich Nietzsche, one of the most influential modern philosophers, introduced the concept of ethical autonomy as a form of self-responsibility to mediate the demands of self-respect, as required by Kant, with the needs of self-love, as explained by Aristotle. Nietzsche believed that with freedom came responsibility for one's own life, shown by his linking of autonomy with freedom and self-responsibility (Nietzsche, 1881).

While these early pioneers in philosophy shaped the modern understanding of autonomy, it was not until the 1970s that two mainstream views on autonomy surfaced within the concepts of free agency and self-determination. In 1970, Gerald Dworkin explored the correlation between an individual acting freely and doing what the individual wants. A year later, Harry Frankfurt explained his view on acting freely, which he uplifts as the human notion of freedom. Both of these important works laid a foundation for relational autonomy theorists, who would spend decades critiquing Dworkin and Frankfurt for neglecting to consider the role of societal structures in realizing individual autonomy. Relational autonomy, which shines light on the social dimensions of individual autonomy and how its realization can be restricted by societal structures, as previously discussed, was introduced in 1989 by Jennifer Nedelsky, bringing with it a multidimensional analysis of autonomy that is better suited for application in reality (Mackenzie, 2021).

More recently, in 2011, a group of theorists introduced the theory of conjunctural action, a new framework that explores the importance of structure and agency in determining human action, based on the belief that "human societies, the product of human agency, are structured through processes that emerge from the social capabilities and propensities of human organisms" (Johnson-Hanks et al., 2011, p. 1). The theory explores how family behavior changes based on the interplay of societal structures and events that reconfigure those structures, concluding that fertility behavior is best understood as a function of both individual traits and social constraints of structures that the behavior occurs within and, therefore, social action cannot be interpreted as a causal effect of structure versus individual agency, cultural influences, or human biology. Rather, human behavior is a product of all four (Johnson-Hanks et al., 2011).

Historically, feminist literature has focused on bodily autonomy and how patriarchal societies seek to control and regulate women's bodies and reproductive capabilities. In *The Second Sex* (1953), Simone de Beauvoir introduces the concept of "the other," by arguing that women are not born but made, and that society constructs femininity as "the other" of masculinity. De Beauvoir introduces the idea of reproductive oppression, or the ways in which patriarchal societies seek to control women's reproductive capabilities through policies, laws, and social norms, and choice feminism, which emphasizes the importance of individual choice and agency in reproductive decision-making. Researcher Elizabeth Wicks agrees, writing, "The principle of autonomy is an important foundational concept for the law of human rights, alongside principles of equality and dignity" (2016, p. 6). Bodily autonomy is indeed an essential ingredient in reproductive freedom. However, some feminist scholars criticize the idea of choice feminism for its sole focus on individual agency and failing to acknowledge the social and economic structures that shape family planning decisions.

While, where there is reproductive justice, there must also be the protection of full bodily autonomy, importantly, the attainment of autonomy does not always equate to the realization of reproductive justice. Full bodily autonomy within the reproductive justice framework requires not only access to comprehensive healthcare but also the ability to secure housing, food, education, and employment, which makes individual choice alone not enough to address the structural inequalities that shape family planning. Today, most feminist theorists recognize the importance of measuring bodily autonomy with a lens of intersectionality. As previously discussed, the concept of intersectionality was introduced in 1989 by Kimberlé Crenshaw. In her iconic work, *Demarginalizing the Intersection of Race and Sex,* Crenshaw argues that the experiences of Black women cannot be understood solely through the lens of race or gender, and advocates for an intersection of race and gender to explain the ways in which marginalized communities face additional barriers to comprehensive healthcare. In 2000, bell hooks further explores intersectionality in *Feminism is for Everybody* by advocating for feminism to be inclusive of all marginalized groups and highlighting how the struggles for racial, gender, and economic justice are interconnected, along with their remedies being interdependent.

Fertility Desires

There are a variety of personal, social, and economic reasons why people choose not to have children, including personal choice, competing priorities with education or career goals, financial considerations, relationship issues, health factors, and environmental concerns. While some may simply prefer not to have children and see it as a personal decision, others may choose to focus on their career or education goals in lieu of starting a family. Still others may choose not to have children due to relationship issues, such as being single or being in a relationship where the other partner does not want children. In addition, health concerns such as genetic disorders or chronic illnesses may lead to a decision to forgo having children, as may financial hardships. Research around fertility desires seeks to understand and address the ways in which individuals, couples, and cultures make decisions about their reproductive health and well-being.

In January 2023, the *National Center for Health Statistics* released data showing birth rates across the United States during the years 1980 through 2007 fluctuated between 65 to 70 births per 1,000 women aged 15 to 44, with birth rates dropping by 20% in 2008 to an average of 55.8 per 1,000 women in 2020 (Martinez & Daniels, 2023). Over the last decade, several studies have investigated the reasoning behind the declining birth rate in the United States, with most linking family planning decisions to the

Great Recession and associated medical, housing, and financial concerns. The years of the Great Recession were also the years when millennials began graduating from college, getting married, and preparing for parenthood. Today, millennials are not only the largest generation in the workforce but also the largest generation of childbearing age.

Studies on fertility desires can be traced back to the 1960s and the work of John Rock, an American obstetrician and gynecologist, who, after observing the physical and emotional toll of unplanned pregnancies on women, developed the first oral contraceptive pill, revolutionizing modern reproductive healthcare (Marsh & Ronner, 2008). Rock's work influenced Ansley Coale, whose 1967 landmark study *The Voluntary Control of Human Fertility* analyzed various methods of contraception available within various historical, social, and cultural contexts and the ability for individuals to control their reproductive choices for each. Coale's work helped to explain drivers of family planning decisions and was influential in highlighting the importance of individual agency and reproductive choice through the context of demographics and population dynamics (Coale, 1967).

In 1980, philosopher Amartya Sen introduced the concept of capability in *Equality of What*, where he argues that governments should be measured against the concrete capabilities of their citizens. Sen believed that only when all societal and economic barriers are removed can individuals truly be said to act out of personal choice and, therefore, it is the responsibility of a society to decide what capabilities individuals are guaranteed (Sen, 1979). That same year, Ronald Rindfuss published a study evidencing that higher education levels of women correlate with lower fertility rates (Rindfuss et al., 1980). What remains unclear, however, is whether a decreased capability for education leads to an increase in the capability to have children, or if an increased capability to have children decreases the capability to receive an education. Further, if higher education inversely correlates with fertility rates, does that mean an increased capability to obtain an advanced degree results in better capabilities to determine reproductive destiny, or does an increased capability to control fertility result in increased capabilities to seek out education, knowledge, skills, and opportunities?

While many highly educated men have spent their careers debating the capabilities and choices of women, few have stopped to consider what might come from asking women what they would choose, if provided access to all their capabilities. In 2011, Martha Nussbaum expanded on

Sen's capabilities approach by emphasizing that all individuals deserve human dignity and the capability to make choices about their own bodies. Nussbaum's work seeks to understand what an individual is able to do and become when provided access to all capabilities, including both the freedom to decide the conditions of one's life and societal opportunities for growth. Nussbaum outlines the conditions necessary for human dignity through life, autonomy, bodily health, emotional health, mental freedom, practical reason, societal belonging, leisure, nature, and environment (Nussbaum, 2011).

Plenty of historical and recent research seeks to explain the gap between fertility desires and birth rates. A 2010 study comparing the number of children an individual claimed to want at ages 16 to 24 with the achievement of their fertility goal by age 40 found that underachievement of fertility goals is likely a consequence of delayed marriage or competing priorities, such as advanced education or career achievements, concluding that discrepancies in the fulfillment of fertility desires are rooted in chances to achieve motherhood (Morgan & Rackin, 2010). In addition, a 2015 study examined how the value assigned to parenthood impacts the achievement of fertility desires and found that the higher the value placed on parenthood, the lower the correlation between fertility intention and social and cultural influence, such as gender norms, religion, family pressure, and career achievement (McQuillan et al., 2015). While social and cultural influences have been linked to fertility desires in many studies, only a few studies consider economic influences, such as income or costs of living, on both fertility desires and the fulfillment of those desires.

The impact of economic recession on family planning decisions is significant. When examining the impact of the Great Depression of the 1930s, the oil shocks of the 1970s, and the economic shocks in Central and Eastern Europe during the 1990s, along with their associated effects on jobs, income, and housing, one study found that unemployment, job instability, and economic uncertainty are strong indicators of swings in fertility levels (Sobotka et al., 2011). Similarly, when a study explored the discrepancy between fertility desires and fertility rates, it concluded that the decline in fertility rates in the United States cannot be explained by changing desires, but rather by barriers to achieving fertility goals, such as student debt and costs of housing and childcare (Stone, 2018).

Along with studies evidencing declining birth rates, recent polls also indicate changes in fertility desires for younger generations. For example, a 2021 poll by *Pew Research* asked Americans about their plans for kids and found, among those who responded, 44% of adults living in the United States indicated no intention of having children, a 7% increase from 2018 (Pew, 2021). A second poll, conducted in 2023 by *Gallup*, found that when American adults were asked about the ideal number of children for a family to have, 45% of respondents favored larger families, defined as three or more children, a 4% increase from 2018 and the highest point since 1971. However, the birth rate across the United States remains low compared with the 1970s, suggesting that Americans' views of the ideal family may not be their personal reality (Gallup, 2023).

The University of Chicago's *National Opinion Research Center* conducted a survey in 2023 on American perspectives of the current state of the economy and found the majority of Americans who responded were greatly concerned about inflation, housing affordability, healthcare costs, and the overall cost of living. Approximately 30% indicated they were experiencing a major financial strain, with the rest experiencing either a minor financial strain or expecting a financial strain to be evident. The same study found 80% of respondents believed the economy was in poor shape and 44% indicated their finances were in worse condition than they expected for their current stage of life. Further, 78% of respondents did not feel confident that life for their children's generation would be better than it has been for them, with 45% believing they did not have a good chance of improving their standard of living (NORC, 2023).

In 2023, sociologists Karen Benjamin Guzzo and Sarah Hayford surveyed groups of individuals born between the years 1981 and 1997 regarding their fertility desires. Respondents' fertility goals, indicated by the number of children they desired during their late teenage years and early 20s, were analyzed against the number of children they had at the time of the survey and then compared to the responses of older generations, or those born before 1980. The study found that younger generations have similar fertility desires as older generations, with a parity of two kids, but the gap between fertility desires and children born has doubled for individuals born after 1980. To explain the widening gap, the researchers shine a light on the economic barriers that younger generations face, concluding that people are choosing to have less children, not because they desire fewer children, but because they cannot afford to have them (Guzzo & Hayford, 2023).

Climate Considerations

In 2022, a group of researchers published findings that Becker's theory of fertility, which explains a positive correlation between income and fertility rates, is no longer valid due to the income-fertility relationship flattening and the education-fertility relationship receding (Hannusch et al., 2022). The findings add complexity to an already complicated collective understanding of what drives fertility behavior. Yet, with the links between education, income, and fertility evolving, factors unaccounted for in previous research may need to be reconsidered. For example, while understanding the fertility intentions of American adults is valuable, the most alarming finding in the 2021 *Pew Research* poll was not the increase in childless households, but that of the respondents who indicated they did not have plans for kids, 56% opted not to provide a reason when asked why.

While economic and social barriers place significant burdens on younger generations and their fertility desires, a lesser-known reason to skip having kids has gained popularity in recent years. Concerns about overpopulation and its impact on the environment have led some people to choose not to have children or to have less children than originally planned. For decades, the IPCC has advocated that reductions in population growth significantly reduce overall carbon demand and mitigate climate change. For example, a 2009 study found that for each child born in the United States, a woman adds about 9441 metric tons of carbon dioxide to her carbon legacy, which is approximately 5.7 times an average woman's lifetime emissions (Murtaugh & Schlax). Accordingly, a 2017 study conducted by two Canadian climate scientists concluded that having one fewer child is the greatest impact an individual can have on reducing greenhouse gas emissions (Wynes & Nicholas, 2017).

This recent phenomenon may be best explained by the *tragedy of the commons,* a theory introduced by Garrett Hardin in 1968 to explain how, in the context of shared resources, when an individual acts solely based on their own short-term interests, which is to take as much of a resource as possible, they are actually acting against the interests of everyone, including themselves, because if everyone were to act solely based on their individual interests, demand for a shared resource would overshadow supply, resulting in the resource eventually becoming unavailable, and society being worse off as a whole (Hardin, 1968). Hardin's theory is an example of the Nash equilibrium, illustrating how everyone suffers when individuals choose to act solely based on their own interests. This may partly explain why a trend has emerged among younger generations to limit their fertility desires in the same way they limit their consumption of carbon-intensive household products and recreational activities.

This camp can be divided into two main categories: those who opt to limit or reduce their lifestyle emissions by having fewer kids and those who choose not to have kids due to uncertainty about what a warming planet means for the prosperity of future generations. For example, a 2018 *Gallup* poll found that 70% of adults aged 18 to 34 were worried about global warming, compared to 56% of adults aged 55 or older. A second survey conducted in 2018 by *The New York Times* found that 33% of the 20- to 45-year-old people surveyed cited climate change as a reason they had or expected to have fewer children than they might have wanted in different circumstances. In addition, a 2022 survey conducted by *The Harris Poll* found that 25% of young adults in America think people should stop having children due to the harm it causes to the environment. The same poll found that 58% of young adults are concerned about the impacts of overpopulation on the planet, with 42% believing future generations' quality of life will be poor (Harris, 2022).

Climate anxiety, defined by the *American Psychological Association* as "a chronic fear of environmental doom," is on the rise among younger generations. A recent study in *The Lancet* found that, in a global poll of 10,000 people aged 16 to 25, 84% of respondents across all countries were at least moderately worried about climate change, with 59% indicating they were very or extremely worried and 75% stating they think the future is frightening. More than 50% said they experience feelings of sadness, anger, guilt, helplessness, and anxiety due to the climate crisis, with almost half saying their feelings negatively impact their daily functioning and 39% indicating they are hesitant to have children due to their climate anxiety (Hickman et al., 2021). Of course, plenty of studies, including this one, also found financial stress and other economic factors, such as childcare and paid family leave, to be significant drivers of fertility behavior. However, the percentage of younger generations citing climate change as an important part of their family planning decisions is steadily growing.

Less understood is whether those who cite economic factors as their reason for deciding to have less kids, or remain childless, understand that, based on their experiences of economic and financial instability, climate change is also influencing their fertility intentions, due to global warming contributing to the increased cost of food, housing, and healthcare and, in turn, to the overall cost of parenting. Many studies have tried to better understand the drivers of fertility behavior, and some have successfully shed light on the factors individuals take into consideration when deciding whether to become a parent or to have another child. In modern times, raising children—from navigating work demands to the dangers of social media to the cost of college—is complicated. Individuals are influenced by a variety of factors, including lived experiences, social influences, cultural values, and partner preferences. However, while fertility preferences cannot be ignored, the focus of the conversation around the global declining birth rate needs to center on access to full bodily autonomy. If we removed all the economic, social, and environmental barriers to becoming a parent of as many children as desired, would birth rates still be on the decline?

References

Coale, A. J. (1967). The voluntary control of human fertility. *Proceedings of the American Philosophical Society, 111*(3), 164–169.

Gallup S. (2023). *Americans' preference for larger families highest since 1971.*

Giddens, A. (1990). *The consequences of modernity.*

Guzzo, K. B., & Hayford, S. R. (2023). Evolving fertility goals and behaviors in current US childbearing cohorts. *Population and Development Review, 49*(1), 7–42.

Hannusch, A., et al. (2022). *The economics of fertility: A new era (no. 17212).* CEPR discussion papers.

Hardin, G. (1968). The tragedy of the commons: The population problem has no technical solution; it requires a fundamental extension in morality. *Science, 162*(3859), 1243–1248.

Harris Poll. (2022). *Personal independence behind declining birth rates.* The Harris Poll.

Hickman, C., et al. (2021). Climate anxiety in children and young people and their beliefs about government responses to climate change: A global survey. *The Lancet Planetary Health, 5*(12), e863–e873.

Johnson-Hanks, J. A., et al. (2011). *Understanding family change and variation: Toward a theory of conjunctural action* (Vol. 5). Springer Science & Business Media.

Kant, I. (1785). *Groundwork of the metaphysics of morals.*

Mackenzie, C. (2021). Relational equality and the debate between externalist and internalist theories of relational autonomy. In *Autonomy and equality* (pp. 32–56). Routledge.

Marsh, M., & Ronner, W. (2008). *The fertility doctor: John Rock and the reproductive revolution.* JHU Press.

Martinez, G. M., & Daniels, K. (2023). Fertility of men and women aged 15–49 in the United States: National survey of family growth, 2015–2019. *National Health Statistics Reports*, (179), 1–22.

McQuillan, J., et al. (2015). The importance of motherhood and fertility intentions among US women. *Sociological Perspectives, 58*(1), 20–35.

Morgan, S. P., & Rackin, H. (2010). The correspondence between fertility intentions and behavior in the United States. *Population and Development Review, 36*(1), 91–118.

Nietzsche, F. (1881). *Nietzsche: Daybreak: Thoughts on the prejudices of morality.* Cambridge University Press.

NORC. (2023). *National opinion on the state of the nation's economy.* National Opinion Research Center. University of Chicago.

Nussbaum, M. C. (2011). *Creating capabilities: The human development approach.*

Pew Survey. (2021). *Growing share of childless adults in U.S. don't expect to ever have children.* Pew Research Center.

Rindfuss, R. R., Bumpass, L., & St. John, C. (1980). Education and fertility: Implications for the roles women occupy. *American Sociological Review*, 431–447.

Sen, A. (1979). *Equality of what?* (Vol. 1).

Sewell, W. H., Jr. (1992). A theory of structure: Duality, agency, and transformation. *American Journal of Sociology, 98*(1), 1–29.

Sewell Jr, W. H., & Sewell, W. H. (2005). *Logics of history: Social theory and social transformation.* University of Chicago Press.

Sobotka, T., et al. (2011). Economic recession and fertility in the developed world. *Population and Development Review, 37*(2), 267–306.

Stone, L. (2018). *Declining fertility in America.* American Enterprise Institute.

Wicks, E. (2016). *The state and the body: Legal regulation of bodily autonomy.* Bloomsbury Publishing.

Wynes, S., & Nicholas, K. A. (2017). The climate mitigation gap: Education and government recommendations miss the most effective individual actions. *Environmental Research Letters, 12*(7), 074024.

CHAPTER 11

Climate Change and Diminishing Prosperity

Reproductive justice recognizes that true reproductive freedom depends on access to the resources needed for healthy child-rearing, not just the ability to make reproductive choices. The right to parent in safe and healthy environments is a crucial aspect of the reproductive justice framework that ensures parents are able to raise their kids in homes and communities with adequate access to clean water, nutritious food, quality healthcare, safe housing, and other essential resources. Indeed, without the right to raise children in safe and healthy environments, in addition to choosing when or if to have children, true prosperity cannot be fully realized.

When people think about prosperity, they often envision healthy, harmonious, and resilient communities with access to what they need to survive and thrive. Some may think of industrial booms or technological advancements. Others may fixate on advanced education, abundant theatre and music, or effective systems of justice. While visualizations and definitions of a prosperous society may vary from person to person depending on lived experiences, culture, and values, the common denominator in each image is the ability of every person to easily procure adequate basic necessities, such as food, housing, and healthcare. There may be many schools of thought on political, economic, and society well-being, but they all agree that without people's basic needs being met, little can be imagined in the space of growth, innovation, or achievement, including

S. Haley Knowles, *Diminished Prosperity*,
https://doi.org/10.1007/978-3-032-07068-5_11

fulfilling one's fertility desires. This chapter discusses the importance of access to adequate housing, food, and healthcare in realizing full bodily autonomy. Adverse impacts on affordable housing, nutritious food, and health outcomes stemming from climate change are reviewed.

Adequate, Nutritious Food

The United States Department of Agriculture (USDA) defines food insecurity as a "household-level economic and social condition of limited or uncertain access to adequate food," while dividing those households experiencing food insecurity into two categories: households experiencing low food security, defined by "reports of reduced quality, variety, or desirability of diet," and those experiencing exceptionally low food security, defined by "reports of multiple indications of disrupted eating patterns and reduced food intake" (USDA, 2025). In 2022, 170 million households in the United States were found to be food insecure, a significant increase from 135 million in 2021 (Rabbitt et al., 2023).

The United States Department of Health and Human Services (UHHS) recently published a report explaining how food insecurity can be experienced temporarily or over long periods of time, while also being influenced by several different economic, social, and political factors. Economic factors, such as unemployment and low wages, contribute to food insecurity by making it more difficult for households to purchase adequate food for all family members. In addition, with high unemployment rates, particularly among low-income populations, barriers to adequate nutritious food increase further. Accordingly, studies show that children living in households with unemployed parents have higher rates of food insecurity than children living in households with parents who are gainfully employed.

Research on contributors to food insecurity in the United States is vast and deep, implicating systemic issues, such as discrimination, education, affordable housing, and transportation, as drivers of increased poverty rates and segregated neighborhoods, which lead to decreased access to adequate nutritious food. For example, disabled adults are at a higher risk for food insecurity due to limited employment opportunities and healthcare-related expenses that reduce income available to purchase nutritious food. Food insecurity can also occur due to a lack of transportation or inadequate public transportation, making it difficult for people to access healthy food options in other neighborhoods. Further, the relationship between transportation and food security tightens when long distances exist between households and full-service grocery stores. Households in urban and rural areas, those closely correlated with low-income neighborhoods with limited transportation, are also likely to be located in food deserts, or areas without access to full-service grocery stores, which results in residents being forced to rely on convenience stores, with higher food prices, lower-quality foods, and less variety of foods, as their main source for groceries (DDHS, 2025).

Plenty of studies evidence how the role of social institutions and structural discrimination significantly influences which households will experience food insecurity, as the households more likely to live in food deserts and lack access to transportation are often those also living with chronic diseases or disabilities, residents of rural areas, and members of racial and ethnicity groups. For example, in 2022, 22.4% of Black households and 20.8% of Hispanic households were found to be food insecure, compared to 9.3% of white households and 10.5% of households nationwide (USDA, 2025). However, while the link between food insecurity and race, ethnicity, and disability status is well-researched and documented, the relationship between food insecurity and social factors is one of correlation, not causation.

Food-insecure households are caused by the inability to obtain adequate, nutritious food, likely caused by income level, not an individual's race or physical ability. For example, in 2022, 36.7% of households determined to be low income were also found to be food insecure, compared to the national average of 12.8% (USDA, 2025). Yet, it cannot be ignored that systemic racism and discrimination create additional barriers to financial gain for some individuals based on their race, ethnicity, and disability status, which means that minority households are discriminated against in the job and housing markets, resulting in decreased levels of income and, therefore, less money to buy food. Importantly, however, is the clarification that being Black or living with a disability is not the reason an individual experiences hunger; hunger is caused by not being able to purchase adequate, nutritious food.

Along with social factors, political factors, such as public policy, globalization, and international conflict, also contribute to food insecurity by determining food availability, accessibility, and affordability. With the consolidation of the food industry from globalization, supply chains are now at greater risk of disruption, leading to higher food prices and associated rates of food insecurity for households globally. For example, in 2019, the number of households determined to be food insecure was 135 million. Within three years, with a jump to 170 million, an additional 35 million households were found to be experiencing food insecurity, likely from consequences from failing crops and supply chain disruptions caused by the war in Ukraine, the aftermath of the COVID-19 pandemic, and climate change, which all contributed to pushing food prices to all-time highs (Rabbitt et al., 2023).

Accessible, Comprehensive Healthcare

Access to affordable, comprehensive healthcare in the United States continues to be a challenge for many communities. In addition, a multitude of economic, political, and social factors and inequalities affect health outcomes, such as how discrimination, racism, lack of affordable housing, lack of education, and limited access to services can contribute to poor health outcomes. Healthcare insecurity can also occur due to geographical factors, such as living in a rural area, which can make it difficult to access healthcare services and treatments, along with decreasing the quality of service obtained.

The health of all communities depends upon not only access to and quality of healthcare but also social determinants of health, which are factors that impact a person's environment and ability to access healthcare, purchase healthy foods, and care for oneself. In 1998, researchers Michael Marmot and Richard Wilkinson, together with the World Health Organization (WHO), established the *Social Determinants of Health*, a framework to explain why discrepancies in health outcomes among populations are better explained by social factors, not by differences in quality and frequency of healthcare. Through their research, Marmot and Wilkinson discovered that the impact of where a person lives and the experiences they have based on their social class is such a strong indicator of health that when a person changes their social and cultural environment, their risk of disease changes. However, the relationship between health and social status is not based solely on income but rather on all the elements that comprise the socioeconomic spectrum. In other words, a person's health status does not determine their position in society; instead, a person's social position determines their health status (Wilkinson & Marmot, 1998).

Included in Wilkinson and Marmot's original framework (1998) is a list of social determinants of health, including stress, social exclusion, social support, work, unemployment, food, addiction, and transportation. Today, *Healthy People 2030,* a campaign funded by DHHS through the Office of Disease Prevention and Health Promotion (ODPHP), organizes the social determinants of health into five domains, namely economic stability, education access and quality, healthcare access and quality, neighborhood and built environment, and social and community context. In addition, health insurance status is a key factor impacting healthcare insecurity, as individuals with chronic health conditions may require more frequent and expensive healthcare services, and lack of adequate insurance coverage contributes to healthcare insecurity, as individuals may not be able to afford the cost of healthcare services and treatments (DDHS, 2025).

However, along with social, political, and economic drivers of health, emerging research illustrates a direct link between hotter temperatures and poorer health outcomes, including worsening malnutrition and increasing risk of infectious disease due to microbes thriving in warmer environments. For example, in 2023, *The Washington Post* teamed up with *CarbonPlan,* a nonprofit investigating the effects of climate change on the modern world, to produce the *Human Limit*, a series of articles on how human populations respond to heat. *CarbonPlan* leveraged wet-bulb globe temperature, which considers air temperature, humidity, radiation, and wind speed using historic heat data applied to NASA's latest climate projections, to measure how heat stresses the human body in 15,500 cities around the world and reported their observations over several weeks (Post, 2023).

The findings are frightening. Among them is evidence of climate-change-induced heat stress causing widespread infectious disease outbreaks, increased malnutrition, rampant homelessness, and uncontrollable migration, with historically marginalized populations suffering the most. For example, in Yemen, where the City of Hodeida is impacted by extreme heat more than any large global city, heat, hunger, and health go hand in hand. Emerging research indicates that children suffer from both the indirect impacts of elevated temperatures, like crop failures that affect their diets, and direct effects, like heat exhaustion, heat stroke, and dehydration. Further, families living in extreme heat and experiencing the lasting impacts of poorer health and decreased nutrition are poised to spiral deeper and deeper into poverty. In the United States, extreme heat is already a leading cause of death, killing an average of 8,500 Americans per year, more than all other weather-related deaths combined. *CarbonPlan* predicts this number to increase to 59,000 by 2050, with the highest-risk areas located in southern California, southwest Texas, and southwest Arizona (Post, 2023).

Safe, Affordable Housing

As every economist knows, the most basic factor that determines housing prices is the balance between supply and demand, which is largely driven by economic stability. When the economy is strong and unemployment is low, people are more likely to have the financial means to buy a home, which can drive-up housing prices due to an increase in demand. Conversely, when the economy is weak and unemployment is high, people are less likely to be able to afford a home, which often lowers housing costs. A similar effect can be found with interest rates that determine how affordable it is to borrow money, which, in turn, drives demand and the resulting impacts to supply.

However, while most changes in the housing market can be explained through supply and demand, what is more complicated is identifying and understanding what is driving changes to economic stability that, in turn, influence supply and demand. For example, other demographic trends, such as population growth, disaster declarations, and migration patterns, impact the housing market as well. When a city or region experiences a rapid increase in population, which could be caused by rural-to-urban migration or a geospatial-specific industry boom, demand for housing increases, which drives up housing prices and decreases the supply of houses for sale, along with decreasing the availability of affordable housing. Further, government policies, such as zoning regulations, taxes, and subsidies, also drive housing prices, along with the segregation of communities, which, in turn, can affect housing affordability for certain populations.

Each year, *CoreLogic* releases its annual Climate Change Catastrophe Reports, which highlight how climate change impacts the housing market from damages caused by wildfires, hurricanes, flooding, and other severe weather events. Recent reports show the increasing frequency and severity of extreme weather events exacerbate the housing crisis by reducing the supply of affordable housing and increasing housing and related costs. For example, in 2022, one in ten homes located in the United States, equating to more than 14 million properties, was damaged by a climate disaster (CoreLogic, 2023). Further, in 2023, when a devastating wildfire tore through the island town of Maui, Hawaii, over 100 lives were lost and more than 2,200 structures were burned to the ground, mostly affordable housing, contributing to over $5.5 billion in property damage and leaving over 7,000 residents in need of shelter. To help combat the pending housing crisis, the Governor of Hawaii signed an emergency proclamation making it illegal to make an unsolicited offer to purchase property in three West Maui zip codes affected by the wildfire (Hawaii, 2023).

While this proclamation may seem odd to some, the aftermath of a climate disaster is often met by property development companies moving quickly to purchase land from vulnerable individuals who find themselves in a financial predicament with their house suddenly destroyed or severely damaged, inadequate insurance coverage, and too little savings to fill the gap. Climate disasters are a catalyst for a new or deepening housing crisis as developers easily snatch up properties abandoned by residents without the means to rebuild, resulting in decreased availability of affordable housing and increased housing costs. For example, in September 2021, *NPR* reported that Sonoma County, California, a high-risk area for wildfires, has seen its median home price rise more than 40% in recent years. Similarly, El Dorado County, California, a small community next to Lake Tahoe known for its wildfires, has seen its median home price rise almost 60% over the last five years. Yet, is the average American family considering climate change as it budgets for rent or mortgage payments?

Recent changes to the insurance industry have added even more complexity to rising housing costs. In July 2023, Farmers Insurance announced it would not renew a third of homeowners' insurance policies in the State of Florida and would limit new policies in California due to worsening climate disasters. The announcement came a couple of months after State Farm Insurance announced it would not renew tens of thousands of policies in California and only one month after Allstate Insurance declared it would no longer provide home or commercial policies in the state due to worsening climate events (Flitter, 2023). In addition, the Deloitte Center for Financial Services found that severe climate-related disasters in the United States escalated by 32% between 2019 and 2022, leading to insured losses rising by nearly 300% over the four-year period (Deloitte, 2024). Further, a 2023 *Policygenius* Home Insurance Pricing Report found that homeowners in several disaster-prone states saw an average

50% increase in housing insurance costs since 2021 (Howard, 2023). In January 2024, Deloitte surveyed 2000 homeowners across 21 states associated with high climate risk and found that dwindling insurance coverage and soaring costs have led to 25% of respondents choosing to go without adequate homeowners' insurance. As more and more homeowners living in high climate-risk regions are unable to secure adequate housing policies, the adverse impacts of extreme weather events on the housing market will only get worse.

At the root of human-caused climate change are greenhouse gas emissions that are fueling global warming, resulting in unpredictable rain patterns, higher temperatures, and more extreme weather events, such as hurricanes and wildfires. At the core of greenhouse gas emissions is consumerism, corporate expansion, and extractive capitalism. In the modern capitalist model, consumer demand drives corporate expansion which, in turn, drives greenhouse gases, climate change, and the resulting adverse impacts on food, housing, and healthcare expenditures. Yet, modern success models do not acknowledge that earning more money and buying more stuff may actually be contributing to the rising cost of living and the declining birth rate. Further, if corporate expansion is driving the depletion of natural resources and contributing to the climate crisis, which, as a result, is driving up the cost of living and making it harder for parents to raise their families, then it can be concluded that the mainstream measurement of economic progress based on limitless market growth is actually at odds with the ability of a society to fully realize prosperity. This leads to the inconvenient truth that unchecked corporate expansion is the largest contributor to global warming, and it can no longer underpin every corporation's year-end ambition.

While many theories have been introduced to address the environmental and social harm associated with a capitalistic system and the idea that expansion is necessary for survival, few are willing to acknowledge the inconvenient reality that boundless economic expansion is not sustainable, and that any economic theory that claims to address the environmental degradation associated with industrial growth must also implicate the societal transformation needed to truly be a sustainable society. In 2022, a group of researchers poignantly proclaimed that the devastating impacts of climate change are more of a consequence of the human-made vulnerability to extreme weather events and less due to uncontrollable climate hazards. Outlining the many ways in which climate disasters are a result of environmental hazards intersecting with vulnerability, such as poor urban planning, a lack of social support programs, or an inadequate healthcare system, the researchers argue that a "conceptual re-orientation is a necessary starting point to identify and leverage structural, systemic, and enabling solutions that transform societies to be more equitable and resilient in the long term" (Raju et al., 2022, p. 2).

Indeed, we need a new form of capitalism grounded in sustainable, responsible practices and a circular economy that seeks to preserve humanity and the planet while not forgoing the value of economic prosperity. We need a new method for balancing supply, demand, and economic governance, while applying the values of sustainability to the social aspects of a capitalist system, one which will provide a new way of thinking about industry growth and long-term viability to encompass the ideals of both capitalism and sustainability without sacrificing the value or compromising the principles of the other. However, to establish a new model of prosperity, we must first acknowledge that historical approaches to designing, developing, and measuring prosperity no longer work for modern society.

References

CoreLogic. (2023). *Climate change catastrophe report.* Retrieved January 4, 2024, from https://www.corelogic.com/category/intelligence/reports/climate-change/

DDHS. (2025). *Social determinants of health. Health people 2030.* https://odphp.health.gov/healthypeople/priorityareas/social-determinants-health

Deloitte Insights. (2024). *Bridging insurance gaps to prepare homeowners for emerging climate change risks.* Retrieved May 10, 2024, from https://www2.deloitte.com/us/en/insights/industry/financial-services/bridging-the-gap-between-homeowners-insurance-companies-climate-change.html

Flitter, A. (2023, July 14). Insurer's retreat in Florida signals crisis with no easy fix. *The New York Times.*

Howard, P. (2023). *Home insurance prices up 21% as homeowners are left to deal with climate change, turbulent market.* 2023 Policygenius Home Insurance Pricing Report. Policygenius.

Office of the Governor of the State of Hawaii. (2023). Seventh Proclamation Relating to Wildfires. https://governor.hawaii.gov/wp-content/uploads/2023/09/2309064.pdf

Rabbitt, M. P., et al. (2023). *Househ old food security in the United States in 2022.*

Raju, E., et al. (2022). Stop blaming the climate for disasters. *Communications Earth & Environment, 3*(1), 1.

USDA. (2025). *Definitions of food security. Food security in the U.S.* https://www.ers.usda.gov/topics/foodnutrition-assistance/food-security-in-the-us/definitions-of-food-security.

Washington Post. (2023). *The human limit.* Retrieved January 3, 2024, from https://washingtonpost.com/climateenvironment/climate-lab/

Wilkinson, R. G., Marmot, M., & World Health Organization. (1998). *The solid facts: Social determinants of health (No. EUR/ICP/CHVD 03 09 01).* WHO Regional Office for Europe.

CHAPTER 12

Moving Away from Traditional Success Models

The relationship between economic development, industrial expansion, and harm to the environment and communities has long been debated by economists, sociologists, and policymakers. However, the correlation between greenhouse gas emissions and societal prosperity is not well understood. Increased knowledge of the relationship between climate change and the cost of housing, food, and healthcare, all of which are critical for maintaining the health of families, communities, and economies, is needed. Further, as climate change increases, impeding access to adequate food, housing, and healthcare, barriers to full bodily autonomy also increase, causing interference in family planning. Yet, how climate change influences fertility desires, and how the impediment of fertility desires infringes upon the health of families, communities, and economies, needs to be explored further. In addition, based on the theory of prosperity, what role does society assign to healthy families, communities, and economies in realizing prosperity? Based on the understanding of the relationship between climate change and healthy families, communities, and economies, what is the role of corporate expansion, as both a driver of climate change and an enabler of economic development, in realizing prosperity? Further, what is the role of public policy in promoting, protecting, and advancing prosperity? The answers to and implications of these questions can no longer be ignored.

S. Haley Knowles, *Diminished Prosperity*, https://doi.org/10.1007/978-3-032-07068-5_12

This chapter discusses the relationship between economic development, carbon emissions, and the resulting change to our climate. The disproportionate impacts of climate change are discussed through the lens of climate equity, along with the evolution of economic theory to better address climate change as a modern externality. To understand the relationship between climate change, the fulfillment of fertility desires, and societal prosperity, we must first understand how and why climate change impacts fertility desires and the pathway through which it does. In addition, to understand the relationship between climate conditions and fertility, we need to explore the drivers of and barriers to fertility behavior. As previously discussed, the cost of raising a child is a motivator for family planning decisions. While an individual may desire to have multiple children, the inability to afford the cost of housing, food, and healthcare, among many other expenses connected to rearing a child, may prevent the individual from fulfilling her fertility desires, which is often a strong measure of societal prosperity throughout the literature. Therefore, if classical economic theory holds that economic progress positively correlates with increased birth rates, can we assume that progress declines along with declining birth rates? Further, if increased costs of living are driving a decline in birth rates, can we conclude that the root cause of rising prices is responsible for diminished prosperity? To explore how we get from the impacts of climate change to fully realizing prosperity, we must first understand how climate change impacts the cost of living.

In 2020, the Commodity Futures Trading Commission (CFTC) published a report on climate risk in the United States financial system, concluding that climate change poses systemic risks to the financial system, affecting multiple sectors, geographies, and assets. The report further concludes that while climate-related financial risks are large, they are unpredictable, as climate change is shifting fundamental environmental parameters, pushing planetary systems to new extremes, and positioning variables such as economic growth, crop yields, and labor productivity to deteriorate quickly and suddenly. Strong, vibrant economies of scale are critical for the prosperity of civilizations, yet the CFTC found that climate change reduces labor capacity and productivity, which, in turn, negatively impacts agricultural yields even more (Martinez-Diaz & Keenan, 2020).

In fact, based on recent reports, climate change is trending toward \$4.65 billion in annual productivity losses for the United States agriculture industry by 2030. Physical impacts on crop yields are predicted to compound with reduced labor productivity, resulting in significant declines in staple crops, such as a 10% decline in corn by 2050. Importantly, along with soy and wheat, corn comprises over 60% of harvested crops in the United States and all three crops are vulnerable to heat, while also being key to growing seasons throughout the United States, which, without adaptation measures, will experience substantial losses from a warming climate. Rising temperatures in key agricultural regions across the United States are even leading some farmers to harvest in the middle of the night to safeguard the quality of their crops, along with the health of their workers (Jägermeyr et al., 2021).

Accordingly, current research shows that without meaningful action to reduce emissions or adapt farming to extreme temperatures, labor productivity losses are expected to reach approximately $500 billion by 2050, which corresponds to about 0.5% of projected GDP by 2030 and 1% by 2050. In other words, in a business-as-usual scenario, the year 2050 could bring projected losses of 0.5% of Gross Value Added (GVA) for 80% of American counties, which comes with a price tag of about $5 million per year for 86% of counties located in the United States. Further, by 2090, total impacts from extreme heat caused by global warming will result in over two billion lost labor hours, which corresponds to $160 billion in lost wages for American households (Martinez-Diaz & Keenan, 2020).

The Relationship Between Economic Development and Emissions

Following the Industrial Revolution, as industrial activity increased and innovation advanced, employment opportunities and household incomes also increased, which resulted in the ability to make goods faster and cheaper, lowering the costs of consumer goods and increasing global consumption. Further, as consumption grew, so did the demand for more household goods, leading to the expansion of industrial development. As manufacturing expanded, so did the reliance on fossil fuels, which led to an explosion of global greenhouse gas emissions, the onset of global warming, and the beginning of anthropogenic climate change.

For decades, humanity's impact on the environment has been evaluated through the lens of an ecological footprint. First referenced in 1998 by Mathis Wackernagel and William Rees in *Our Ecological Footprint: Reducing Human Impact on the Earth*, an ecological footprint measures what is required of the planet to sustain an individual, a family, a community, or an economy based on the number of natural resources relied upon for the goods and services that comprise each lifestyle. In other words, an ecological footprint evaluates population and consumption against the capacity of nature. The model is commonly used to analyze consumption per capita based on four categories: energy, built environment, food, and forestry products. Based on this measurement, at the turn of the millennium, research suggested that humans had already overshot the planet's capacity by 20%, with humanity's collective ecological footprint expanding 80% during the last 40 years of the twentieth century, mostly due to the industrial revolution. Further, based on this model, the United States has the greatest environmental impact of all countries, responsible for about a fourth of global emissions caused by human activity with only about 4% of the world's population (Hammond, 2006).

Yet, the idea that economic development and associated industrial activity, represented by increased emissions, negatively correlates with the health of a society is a recent revelation, not an accepted truth by classical economists. Conversely, during the 1950s, Simon Kuznets introduced the idea of the Kuznets curve, which illustrates how, as economic development increases, inequality first increases and then decreases as market forces redistribute income. While many studies have disproven this theory, showing that there is no evidence of inequality reversing as economic growth continues, the curve was repurposed in 1991 to apply a similar theory to environmental degradation. The Environmental Kuznets Curve, known as EKC, theorizes that, again, as economic development increases, environmental harm first increases but then decreases due to market forces placing higher value on environmental attributes, such as clean air, clean water, and temperate climate conditions. The main inference from the 1991 study is that, since increased economic development will at some point decrease environmental harm, any effort to limit economic growth will have the opposite effect on the environment, meaning that by reducing economic development, environmental harm will eventually increase more (Mazurek, 2011).

However, in 2019, after analyzing data from a 32-year period in 160 countries, two researchers discovered a positive correlation between economic growth and carbon emissions, but only up to a certain point, when economic growth, measured by GDP per capita, reverses course and begins to decrease as emissions continue (Wang & Li, 2019). Building on these findings, a second group of researchers in 2023 examined data from the ten countries with the highest rates of emissions to better understand the relationship between long-term economic growth and carbon emissions. The group had similar findings, concluding that in order for economic growth to be sustained, the global economy must transition away from fossil fuels toward carbon neutrality by embracing renewable energy sources (Khan et al., 2023).

With decades of research evidencing the growing inability of the planet to sustain human life measured by modern industrial activity and consumer behavior, most economists and policymakers disagree about what, if any, changes to economic development are required. A study in 2012 by researchers Cameron Hepburn and Alex Bowen explored this question by analyzing the viability of three schools of thought on environmental limits to economic growth, organized by those who believe growth is boundless, those who think natural resource depletion will place conditions on growth, and those who argue that growth cannot continue indefinitely. Ultimately, the researchers conclude that continued growth is feasible, but not without a decoupling of economic development and adverse impacts to the planet, highlighting the belief that technological advances through an expansion of the intellectual economy will bring about the change needed by the year 2050 (Bowen & Hepburn, 2012).

While solutions backed by time and advances in technology are difficult to quantify, the need for economic growth to distinguish itself from environmental harm is easy to measure. In 2023, the IPCC concluded that four industries are responsible for 80% of global emissions, namely power generation, industry, transport, and buildings, with agriculture and forestry responsible for the remaining 22%. The findings build upon a 2021 report that links 100 fossil fuel companies to 71% of global industrial emissions since the IPCC formally recognized human-induced climate change in 1988. The same report found that, since that time, over 50% of global industrial emissions have been caused by only 25 companies (Megura & Gunderson, 2022). While technology advancements bring hope for emission abatement and capture, emerging research suggests that technology alone will not solve the climate crisis, which is a deeply rooted systemic problem that requires transformational systemic solutions.

Climate Equity

With all the reports, studies, and unequivocal scientific evidence that human-caused climate change is happening, it is getting worse each year with increased adverse impacts on adequate, affordable food, housing, and healthcare, and its consequences threaten the very livelihood of families, communities, and societies. Why then is more not being done to address the climate crisis? An important part of the conversation around climate change is understanding the difference between the populations who are bearing the consequences of a changing climate and those who have the power to make public policy and private industry decisions to influence the rate of climate change and its associated adverse impacts. As such, the speed (or crawl) to market with climate solutions may be best explained by the (unequal) distribution of climate impacts.

As previously discussed, the consequences of a warming planet, such as extreme flooding, wildfires, failing crops, and infectious disease, are felt more strongly by low-income communities of color, particularly women and single mothers, those already at risk of food, housing, and healthcare insecurity. For example, a 2023 study found that systemic historic and persistent racist policies push Black and Latinx households into areas more prone to flooding, with higher levels of pollutants, and with worse housing conditions, making these households more susceptible to climate change effects. The study analyzed the District of Columbia's housing market and found that climate change crises disproportionately affect renters, residents with low incomes, and Black and Latinx households, even though less dense neighborhoods and wealthier households contribute higher levels of greenhouse gas emissions on average. In addition, Black and Latinx households and those with lower incomes are also more likely to be energy insecure and unable to afford proper heating and cooling for their homes (Burton, 2023).

Climate equity, or climate justice, seeks to address the ways marginalized communities are disproportionately affected by the impacts of climate change and the lack of access to resources and opportunities to adapt and mitigate these impacts. The concept is also not new. In fact, researchers, community leaders, and scientists have pointed out the unequal distribution of climate impacts for decades. In 2002, a group of environmental justice advocates came together to publish the Bali Principles of Climate Justice, which focus on four main canons to: (1) reject the "growth-based model of economic development that has produced the climate crisis"; (2) denounce the unequal distribution of the consequences of our reliance on fossil fuels which results in the disproportionate adverse impacts on historically marginalized communities; (3) pursue procedural and participatory justice, which advocates for direct participation in the decision-making process by those deemed to be most affected by the outcome; and (4) call for corrective justice "through the transfer of resources from those with historic responsibility for climate change to those most susceptible to its consequences" (Gonzalez, 2021, p. 113).

The climate justice movement shines light on the climate crisis as fundamentally a crisis of humanity driven by the unequal distribution of power and consequence, and systemic issues like sexism, racism, and classism that both create and sustain global warming and the disproportionate impacts on historically marginalized communities or those with lower socioeconomic power. The movement emerged in the late 1990s following the recognition that those least responsible for the emissions driving climate change, and those least equipped to adapt to a changing climate, would be those to suffer the most from climate change, mostly women of color (Dolšak & Prakash, 2022). In addition, the energy justice movement, which seeks equity in the social and economic participation in the energy system while also remediating the harms caused by the energy system to historically marginalized communities, emphasizes a just transition to clean, low-carbon energy sources that are accessible, affordable, and managed democratically for all communities (McCauley & Heffron, 2018).

An Evolution of Economic Theory

In 1949, French political economist Georges Bataille introduced a new economic theory called the *general economy theory*, which revolves around Bataille's approach to consumption and his concept of the *accursed share*. In a departure from classical economic theorists motivated by restriction and scarcity, Bataille believed that in all economies exists an accursed share, or an excess of energy that can be leveraged for productive growth, through consumption of art, culture, or other leisure activities deemed acceptable by a society, or spent in lavish, gluttonous, unproductive means, such as war. Regardless of how the excess energy is spent, Bataille argues that the accursed share is always destined for waste within the general economy, but how this excess energy takes shape informs the understanding of "luxury" within a society. Most importantly, the concept of excess energy was developed by Bataille following his reflection on the indigenous tradition of the potlatch, a celebratory gift-giving custom practiced by several First Nations and also known as the revered Lakota Giveaway ceremony.

Understanding how a society translates excess into luxury is an important measure of how "good" the society is, or, based on the views of

Joseph Berliner (1999), an economy's arrangement of citizens as consumers and as workers, and the balance between the distribution of property and the opportunity for human development. Berliner believes that the failure of all socialist economies is inevitable due to the absence of free markets, and that a sustainable good society can only be established as a properly managed market driven by competition and governed by democracy (Berliner, 1999). Interestingly, Berliner does not include "growth" as an essential element in his vision of a good society, which begs the question: can a good society be established by a competitive market and sustained without growth if consumer preference changes?

In 2011, theorist Shann Turnbull, in response to his own plea that society desperately needs a new kind of market economy, one that is not only economically prosperous but also environmentally responsible and societally good, offered *ecological capitalism* by pointing out that the modern understanding of a sustainable global good society is at odds with traditional economic policies, those which require full employment of the labor classes and boundless extraction of natural resources. First introduced in Hardin's *Tragedy of the Commons,* the idea of a new economic system that achieves prosperity without growth is a compelling one. However, the addition of "capitalism" in the name seems to step away from Hardin's original premise, as does the concept of *sustainable capitalism,* introduced by Al Gore, former Vice President of the United States, and David Blood, an experienced investment banker, in their publication, *A Manifesto for Sustainable Capitalism,* and through the founding of their new sustainable investment firm, Generation Investment Management (Gore & Blood, 2010). Sustainable capitalism is presented as a new form of capitalism that integrates environmental, social, and governance (ESG) considerations into a firm's decision-making and aspires toward sustainable growth.

However, shortly following the new venture's announcement, John Fullerton, an experienced impact investor, took to his blog to decry the idea of "sustainable growth" as an oxymoron. Fullerton argues that "despite all the advances in technology leading to resource productivity improvements, and assurances from many otherwise intelligent economists, there is no evidence of a decoupling between GDP growth in the aggregate and the absolute material throughput of the economic system" (2010). Instead, Fullerton advocates for *regenerative capitalism,* or the need to look beyond net-zero emissions to a net-positive impact on the planet. In his 2015 publication, Fullerton introduces the eight principles of regenerative capitalism, namely: (1) the human economy is embedded within natural ecosystems; (2) the measurement of wealth goes beyond what can be reduced to money; (3) innovation and adaptation are key to societal health; (4) economic prosperity depends on the empowered participation of those impacted; (5) culture, traditions, and histories both form and nurture a healthy economy; (6) the edges of systems hold the greatest value for creativity, innovation, and transformation; (7) a healthy material economy is also a robust circular economy; and (8) balance is the key to systemic health.

Fullerton grounds his theory of regenerative capitalism in the idea of a *post-growth economy,* first introduced in 2009 by James Speth, an American policymaker. According to Speth, in a post-growth economy, families, communities, and the environment will not be sacrificed for private sector growth in the name of progress. Instead, the idea of boundless economic growth, normalized as an inevitable, necessary part of progress and an excuse for not addressing long-term social needs, will cease to have a following. Rather, in a post-growth society, the focus will be less on consumption, market capitalization, and maximizing shareholder value, and more on addressing the needs of families, communities, the environment, and all of society to improve the overall quality of life for all its citizens (Speth, 2009).

The idea that boundless economic growth is not sustainable may best be explained by the work of Johan Rockström (2009), a Swedish scientist who introduced a framework to measure the preconditions for human development through a set of nine interlinked planetary boundaries that define the safe operating space for humanity with respect to the system of Earth and its subsystems and associated ecosystem services. Rockstrom argues that, by crossing the boundaries, systems of the planet shift with potentially disastrous consequences for humanity. As of 2023, six of the nine boundaries have been crossed, namely climate change, change in biosphere integrity, land-system change, altered biogeochemical cycles, introduction of novel entities, and freshwater change, with three planetary boundaries remaining intact but in imminent danger of being crossed: ocean acidification, atmospheric aerosol loading, and stratospheric ozone depletion (Rockström et al., 2009).

Building upon Speth's and Rockstrom's theories, economist Kate Raworth (2017) introduced *Doughnut Economics* as an alternative to classical market-focused economics models. Instead, Raworth's model measures progress as a balance between meeting basic human needs, defined as the social foundation derived from the SDGs, and honoring Earth's ecological limits, defined as the ecological ceiling formed by the nine planetary boundaries. Complementary to Raworth's model are Smith's *invisible hand*, Bataille's *excess of energy*, and Nietzsche's *test of the eternal return* that may offer an alternate path, one more ethical and sustainable, for measuring prosperity (Retsikas & Marsden, 2018). For example, a recent 2023 study analyzed responses from the governments of the 38 Organization for Economic Cooperation and Development (OECD) countries following extreme weather events and found that in countries with high levels of economic activity, an increase in harm, or the perceived threat of harm, caused by climate disasters is positively associated with an increase in innovation leading to sustainable development (Ma et al., 2023). Yet, how much closer to the SDGs would the world be with a collective understanding of what truly benefits the public good, how overconsumption associated with luxurious lifestyles is harmful to the greater good, and why the actions of society today will live on in perpetuity for the good (or detriment) of future generations?

While some may derive hope from the results of emerging research, the findings confirm that drivers of change continue to be tied to indicators of productivity and capital gain, which increasingly "treat as a derivate the things we actually value, including not only economic wealth and the distribution of income, but also other dimensions of prosperity: social solidarity and participation, personal agency and opportunities, and environmental sustainability" (Dirksen et al., 2022). However, even within this perspective, there continues to be a polarization between what is good for business and what is good for society. In 2021, economics professor Alex Edmans challenges the traditional view that value generated by capitalism is a fixed pie in his book *Grow the Pie*. Edmans argues that any slice given to stakeholders at the benefit of society is not at the expense of shareholders or profit, and vice versa. Instead, by investing in society, a company can grow the pie, which ultimately benefits everyone, including shareholders, in the long term.

More promising is a new theory of sustainable growth, also introduced in 2021, that explores the relationship between fertility desires, resources, and technology to create pathways toward an economic model that is no longer dependent upon economic inputs in the form of natural resources. The theory, formulated by economist Pietro Peretto, outlines three phases of economic conditions that lead to an evolution of natural resource extraction toward technological advancement to transform economic growth measured by the accumulation of physical assets. Instead, prosperity is measured based on the accumulation of knowledge as the primary driver of economic growth, compelled by not only technological advancement but also deeply rooted institutional transformation (Peretto, 2021).

While many economists and scientists disagree on the best new approach to measure economic growth, and regardless of which new economic theory resonates the best, what is clear is that traditional approaches to economic models no longer work for our society. With significant evidence that climate change is positively correlated with factors contributing to significant financial strain on American households, such as increased destruction to property, higher expenditures on groceries, and added healthcare costs from heat-related medical needs, the implications of a changing climate for society are clear. In the era of climate change, traditional approaches to economic models fall short and are both creating and sustaining reproductive injustices for families and communities across the country.

References

Berliner, J. S. (1999). *The economics of the good society: The variety of economic arrangements*. Blackwell Publishers.

Bowen, A., & Hepburn, C. J. (2012). *Prosperity with growth: Economic growth, climate change and environmental limits*. Working paper, (93).

Burton, E. (2023). Addressing climate change through housing policy: Lessons from the Greater DC Region. *The Washington Post*.

Dirksen, J., et al. (2022). *Towards economic and social prosperity measurement beyond GDP*.

Dolšak, N., & Prakash, A. (2022). Three faces of climate justice. *Annual Review of Political Science, 25*, 283–301.

Fullerton, J. (2010, June 28). *Is sustainable capitalism an oxymoron?* Capital Institute.

Gonzalez, C. G. (2021). Racial capitalism, climate justice, and climate displacement. In *Oñati socio-legal series, symposium on climate justice in the Anthropocene* (Vol. 11, pp. 108–147).

Gore, A., & Blood, D. (2010). Toward sustainable capitalism. *The Wall Street Journal*.

Hammond, G. P. (2006). People, planet and prosperity: The determinants of humanity's environmental footprint. In *Natural resources forum* (Vol. 30, pp. 27–36). Blackwell Publishing Ltd.

Jägermeyr, J., et al. (2021). Climate impacts on global agriculture emerge earlier in new generation of climate and crop models. *Nature Food, 2*(11), 873–885.

Khan, I., et al. (2023). Striving for carbon neutrality and economic prosperity in the top ten emitting countries: Testing N shape Kuznets curve hypothesis. *Journal of Cleaner Production, 429*, 139641.

Ma, Y., et al. (2023). From traditional innovation to green innovation: How an occurrence of natural disasters influences sustainable development? *Sustainable Development.*

Martinez-Diaz, L., & Keenan, J. M. (Eds.). (2020). *Managing climate risk in the US financial system.* US Commodity Futures Trading Commission.

Mazurek, J. (2011). Environmental Kuznets curve: A tie between environmental quality and economic prosperity. *Economics and Management.*

McCauley, D., & Heffron, R. (2018). Just transition: Integrating climate, energy, and environmental justice. *Energy Policy, 119*, 1–7.

Megura, M., & Gunderson, R. (2022). Better poison is the cure? Critically examining fossil fuel companies, climate change framing, and corporate sustainability reports. *Energy Research & Social Science, 85*, 102388.

Peretto, P. F. (2021). Through scarcity to prosperity: Toward a theory of sustainable growth. *Journal of Monetary Economics, 117*, 243–257.

Raworth, K. (2017). *Doughnut economics: Seven ways to think like a 21st-century economist.* Chelsea Green Publishing.

Retsikas, K., & Marsden, M. (2018). Alternate modes of prosperity. *HAU: Journal of Ethnographic Theory, 8*(3), 596–609.

Rockström, J., et al. (2009). A safe operating space for humanity. *Nature, 461*(7263), 472–475.

Speth, J. G. (2009). From crisis to sustainability. In L. Mazur (Ed.), *A pivotal moment: Population, justice, and the environmental challenge* (pp. 329–338). Island Press.

Wang, Z. X., & Li, Q. (2019). Modelling the nonlinear relationship between CO2 emissions and economic growth using a PSO algorithm based grey Verhulst model. *Journal of Cleaner Production, 207*, 214–224.

PART IV

A New Model of Prosperity: Rethinking Success Models in the Era of Climate Change

CHAPTER 13

A New Measure of Prosperity

To protect the ability of families and communities to freely choose whether to have a child and the ability to parent in safe and healthy environments, we must acknowledge and remediate the primary driver of climate change: unchecked corporate expansion disguised as economic progress. We must find a new approach for determining and measuring the success of a company, one that is grounded in sustainable, responsible growth, while also responding to market conditions and the needs of society. In the era of climate change, traditional approaches to economic models fall short. Building upon what we learned through Raworth's Donut Economics and Rockstrom's planetary boundaries, the set of success criteria used by both public and private enterprise requires another look. While Raworth and Rockstrom establish environmental and social considerations and planetary limits, their theories lack a practical and pragmatic approach to corporate strategic planning that considers economic growth alongside environmental and human sustainability based on current market conditions. This chapter introduces a new model of prosperity by combining the principles of capitalism and sustainability to enable responsible, sustainable economic growth. Under the new model, in order for a company to expand, it must show that corporate expansion is meeting a societal need economically, transparently, sustainably, and responsibly. A new model of prosperity provides an opportunity to promote and protect environmental, social, and economic progress.

S. Haley Knowles, *Diminished Prosperity*,
https://doi.org/10.1007/978-3-032-07068-5_13

A New Prosperity Model

The idea that corporate expansion must be restricted simply for the sake of restricting growth is not practical, but neither is advocating for corporate expansion just for the sake of expansion. Therefore, we need a better model for understanding when, why, and how corporate expansion contributes to the overall good of society and when it hinders it. There are many ways to measure success, and not all of them fuel climate change and increase the cost of food, housing, and healthcare for families and communities. While there will always be a need for corporate expansion, now is the time to put forth a better method to determine when that need exists, along with the factors required to make the determination. By evolving to a new model to measure the success of a company, we can not only better respond to and meet the needs of markets, but we can also protect the well-being of communities and promote the health of the environment, all of which contribute to societal prosperity.

I offer a new model of prosperity, building upon the work, research, and teachings of many economists, sociologists, and community leaders, and leveraging my own experiences working in Corporate America, community engagement, and environmental sustainability, to measure whether market expansion meets the preconditions needed to promote, enable, and sustain healthy families, communities, and economies with four market approaches—(1) engage, (2) evolve, (3) enhance, and (4) expand—through four market preconditions: (1) meets an immediate societal need economically, (2) considers long-term implications transparently, (3) manages capital inputs and outputs responsibly, and (4) establishes a value chain sustainably. Under a new model of prosperity, corporate expansion will be more responsible and more sustainable, giving way to remediating the climate crisis, controlling the rising cost of living, and increasing the likelihood that individuals and families can fulfill their fertility desires.

Based on the new model, before a company can plan for market expansion, it must consider the long-term impacts of such growth transparently, which requires the company to first confirm that its expansion in the market is not only economical, but also responsible and sustainable by (1) engaging with stakeholders to identify and understand societal needs and expectations, (2) evolving corporate practices to be rooted in responsible research, design, and innovation, and (3) enhancing circularity processes to eliminate inefficiencies and unnecessary waste. Figure 13.1 illustrates the new prosperity model's four market approaches—engage, evolve, enhance, and expand—and the overall goals and benefits of each.

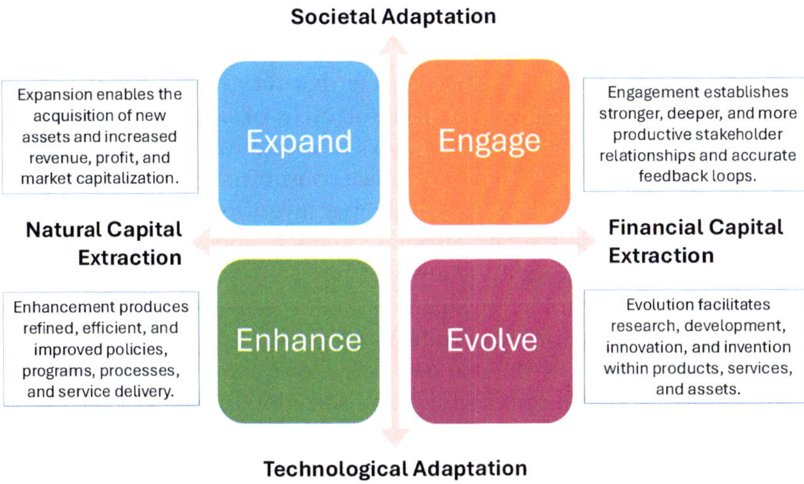

Fig. 13.1 Rethinking market approach

Engage

The first market condition in the new prosperity model is engagement. Engagement helps a company understand how it can best create a business model around meeting an immediate societal need economically. In order to expand responsibly and sustainably in an industry or market, a company must first engage with stakeholders to identify and understand the needs and expectations of all its stakeholders and of society. Too many organizations forgo proper stakeholder engagement, overestimating that the company's business model meets the current problems and desires of society, which can change or evolve quickly. As a result, the company risks misinterpreting market needs, which can lead to wasteful, unnecessary production of goods and services that do not meet the current needs of the company's targeted customers or that do not effectively address what those needs are. Without ongoing, proactive stakeholder engagement, a company does not have a deep understanding of the needs of society and, therefore, cannot design its business model around solving societal problems.

Other companies first decide what their product is and then spend millions of dollars each year in marketing campaigns aimed at convincing consumers that they need the products or services that the company is selling. In traditional growth models, the measure of success is not on creating the perfect product to solve a societal problem; rather, it is on how well a company can convince society that they have a problem that needs solving. This results in a harmful cycle of companies spending money (and emitting greenhouse gases) to create unnecessary products that they then ferociously seek to convince consumers to buy in their pursuit of a profit to recoup their costs and repay investors. What if companies designed their business model around products and services that they already knew society needed—and would buy?

With proper stakeholder engagement, a company can better tailor its business model and associated products and services to both the needs of society and the needs of the market. This allows a company to not only operate sustainably and responsibly, but also economically, as the company's products and services will likely be highly marketable. The engagement market condition requires an assessment of whether the company truly understands the needs of society and directs the company to engage with stakeholders if the assessment reveals the company's business model is not designed to solve a current societal problem. To engage with stakeholders effectively, companies can leverage the Sustainable Development Goals (SDGs), adopted by the United Nations in 2015 to create "peace and prosperity for people and the planet." The SDGs are a roadmap for addressing the most pressing societal needs of the planet and provide insight into how to engage around the development goals in a way that can benefit both a company's bottom line and the needs of society.

Evolve

After a company meets the first market condition and confirms its business model is designed to meet a current societal need economically, the second market condition confirms the company's corporate practices are rooted in responsible research, design, and innovation. Although layers of contractual relationships and processes can muddy perspectives, corporate value chains are ripe with unethical practices and inhumane labor conditions. In historical growth models, companies focus primarily on the bottom line and seldom assess their full value chain to identify and remediate unethical practices associated with both their direct and indirect operations. While this can save money in the short term, the expense associated with perpetuating or being linked to human rights violations often proves to be much more costly over the long term. Instead, the new prosperity model requires companies to proactively minimize the adverse impacts of their operations, supply chains, and business relationships before they expand their operations in the market.

The evolution market condition also provides an opportunity for a company that redesigned or upgraded its business model during the first market condition to evolve its practices to align with its new or improved business model, resulting in responsible business practices designed to work for its sustainable business model. At the heart of responsible business practices is environmental and social due diligence. While some industries have a higher risk of adverse impacts to communities and the planet, all companies must have a due diligence program to identify and manage the environmental and social risks and adverse impacts associated with their business model. In order to evolve corporate practices properly, companies can look to international industry standards for environmental and social due diligence, including OECD Due Diligence Guidance for Responsible Business Conduct and the Guiding Principles on Business and Human Rights (UNGPs). These standards provide a roadmap for implementing a robust, comprehensive due diligence model designed around the needs of the company. While getting started on this journey can be overwhelming, the goal is not to implement a perfect process overnight. The goal is to get started—to do something today that puts the company on the right path to managing the inputs and outputs of its value chain responsibly, and then to take another similar action tomorrow.

Enhance

The third market condition confirms that a company's processes are efficient and environmentally sustainable. In order to grow responsibly and sustainably in a market, a company must first enhance its circularity processes to eliminate inefficiencies and unnecessary waste. In the historical growth model, while companies eliminate inefficiencies as best they can, they rarely prioritize environmental sustainability over cost savings and convenience. Raw material procurement, fuel sources, building renovations, data centers, office supplies, and vehicle usage are all areas where sustainable improvements can enhance a company's ability to expand responsibly and sustainably in a market or industry. Often, enhancing company processes, including improved procurement activities, product and packaging design and development, and service delivery, requires investing in improvements upfront that will pay off over time. For companies measuring success based solely on historical growth models, a short-term mindset often discourages investments that cannot be realized quickly. However, in order for a company to expand responsibly and sustainability in a market or industry, shifting toward a long-term growth mindset is absolutely necessary.

Sustainable, environmentally efficient processes are vital for sustainable, responsible growth, including measuring the greenhouse gas emissions of a company's value chain. A company cannot manage what it does not measure. The foundation of all environmentally sustainable programs and initiatives begins with measuring the company's impact on the environment. To enhance processes efficiently, companies can deploy the guidance of the Greenhouse Gas Protocol to measure and report on their carbon footprint. In addition, companies can leverage the Taskforce on Nature-related Financial Disclosures (TNFD)'s LEAP assessment to evaluate and report on the nature-related risks and opportunities throughout their value chain. Once a company understands its baseline environmental impact, it can then design a strategic plan to reduce, replace, and eliminate inefficiencies and environment harm, as appropriate.

For companies that design and manufacture consumer goods, conducting a life cycle assessment on their products is an effective way to gain insights into areas where processes, including materials, can be enhanced to improve environmental sustainability. Further, aligning product and service life cycles with the circular economy provides many benefits to a company. Based on the company's business model, a circular economy aims to keep materials and products in use by consumers and manufacturers for as long as possible. Companies can minimize waste and reduce resource consumption by closing loops in energy flows through enhanced processes that reuse, repair, refurbish, and recycle materials, ultimately leading to significant cost savings, along with enhanced, sustainable processes.

Expand

When a company has met the engage, evolve, and enhance market thresholds, a company has comprehensively and transparently considered both the short-term and the long-term implications of its growth. The company's business model meets an immediate societal need economically through its products and services. The company also has responsible business practices around its approach to procurement, market research, product design, and service delivery that minimize any adverse impacts to communities and society that may stem from its business operations. Further, the company has efficient, sustainable processes that minimize waste and resource consumption and adverse impacts to the environment. Based on the new prosperity model, when a company has satisfied the first three market conditions, a company can sustainably and responsibly expand in a market or industry, which results in market growth that is not only beneficial to the company but is also beneficial to the prosperity of the planet and of society.

Under the new model, in order for a company to expand, it must show that corporate expansion is meeting a societal need economically, transparently, sustainably, and responsibly. In situations where a company cannot demonstrate that all four market conditions exist, the company has a roadmap for pursuing success through other means, which not only allows it to maintain and, at times, even add firm value but also puts the company on the path toward meeting the market conditions necessary for market growth. The new model illustrates how market expansion is only one way for a company to demonstrate its value and how stakeholder engagement, research and development, and improved operational efficiency are examples of success criteria that are often undervalued, overlooked, and misunderstood.

The new model of prosperity embraces the principles of sustainability without sacrificing those of capitalism. As a result, the model evolves extractive capitalism to incorporate the principles of sustainability in a way that is practical, actionable, and measurable. Through the four market conditions—engage, evolve, enhance, and expand—a corporation is able to reap the benefits of capitalism without causing harm to society and the planet. By satisfying all market conditions, a company is positioned for responsible, sustainable expansion of its mission to solve societal problems without also creating societal problems. When executed properly, the new prosperity model may help mitigate the carbon emissions stemming from corporate operations and expansion and, in return, may lessen the impact of climate change on the cost of living.

Figure 13.2 illustrates the four market preconditions required for corporate expansion and the alternative market approaches—engage, evolve, and enhance—for situations when corporate expansion is determined at the time to be irresponsible, unsustainable, or both. Each alternative market approach provides a corporation with alternative measures of success

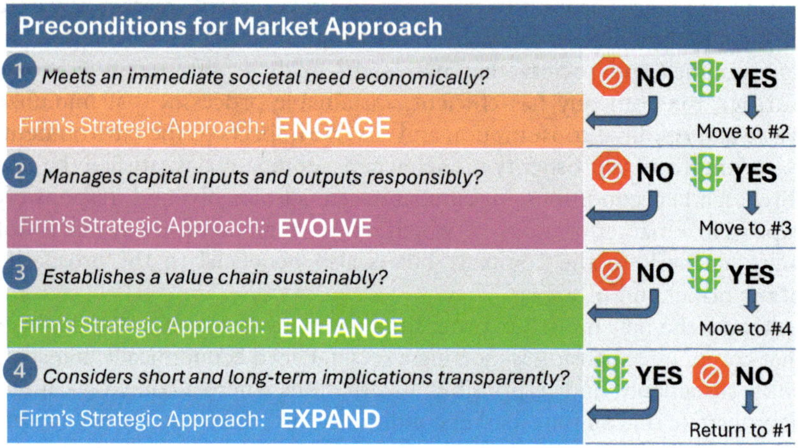

Fig. 13.2 Preconditions for market approach

while also putting it on the pathway toward achieving the conditions necessary for responsible, sustainable growth.

To accomplish this, the sustainability and corporate social responsibility practices of corporations must pivot away from yearly, unchecked goals of "sustainable growth" endeavors, peppered with well-meaning philanthropic campaigns, and shift toward authentic, regimented risk and impact assessments that measure and evaluate whether the growth of assets within a particular market over a specified time frame is sustainable and responsible, not only good for the bottom line. For sustainable growth to be meaningful and impactful, companies must start asking the right questions to pinpoint exactly where their business model sits within current market conditions, so they can properly and responsibly gauge where they are on the maturity spectrum for sustainable growth, and what the appropriate next steps are to reach full maturity, or, in other words, a fully sustainable business model able to responsibly grow in the market. This requires preproduction intervention and due diligence to measure and evaluate current market conditions and inform corporate decision-making.

Figure 13.3 outlines the sustainable growth maturity process and associated challenges and benefits of each stage—from engaging with stakeholders to understand societal needs, to evolving current practices to remediate the risk of human rights violations, and to enhancing operations to reduce inefficiencies.

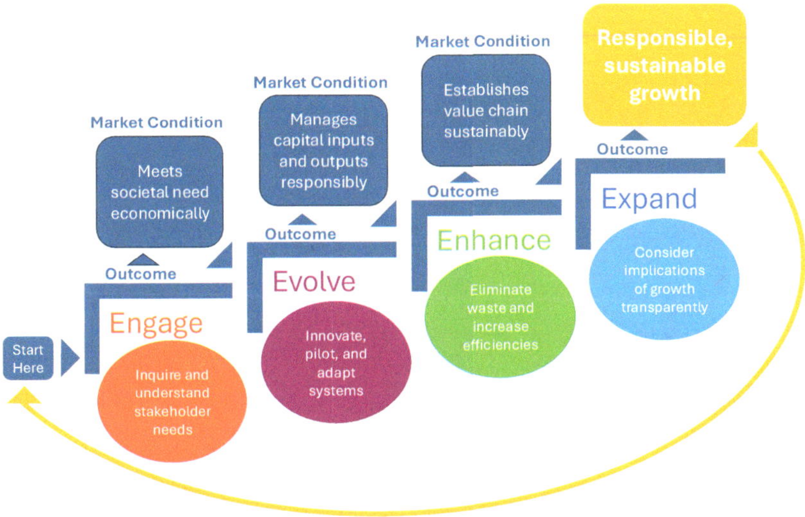

Fig. 13.3 The sustainable growth maturity spectrum

A new prosperity model shifts economic growth models from the narrow focus on extractive capitalism and shareholder maximization at any cost to other stakeholders, including American households, toward a more equitable and sustainable growth model that benefits all stakeholders, including families, communities, and economies. While the market conditions may appear to overly burden companies in the short term, the benefits of the new prosperity model extend beyond a company's stakeholders to the company itself. By shifting toward a more sustainable and responsible business model, a company can better manage risk and contribute to its valuation and long-term viability in the market. However, for private industry to fully realize the value of a new prosperity model, corporations must first shift their policies, practices, and mindsets from ones rooted in short-term gains toward ones centered around long-term responsible and sustainable growth.

CHAPTER 14

Private Industry Considerations

In January 2024, the World Economic Forum (WEF) released its Global Risks Report, an annual report that explores the most severe risks the world may face over the next decade. The 2024 report assesses risk against a backdrop of rapid technological change, economic uncertainty, a warming planet, and conflict, and calls for a new approach to economic growth that balances efficiency with long-term sustainability and equity, examining speed and quality together (WEF, 2024). Similarly, a United Nations report published in February 2024 concludes that the global extraction of raw materials is expected to increase by 60% by the year 2060, building upon findings that natural resource extraction has soared by 400% since 1970 due to industrialization, urbanization, and population growth. Further, the report illustrates how the extraction of natural materials is responsible for 60% of global heating impacts, 40% of air pollution impacts, and 90% of global water stress (UN, 2024). However, what is less addressed is the relationship between private enterprise and natural resource extraction, particularly how corporations' annual strategic growth goals contribute to natural capital depletion and global warming.

While the recent economic reports and shocking statistics may seem dire to some, what is more scandalous is that the same publications have been reporting on these risks for years, even decades. The idea that corporate expansion and unchecked market growth are associated with adverse environmental and social impacts is not a recent revelation. In fact, many

S. Haley Knowles, *Diminished Prosperity*,
https://doi.org/10.1007/978-3-032-07068-5_14

companies publish annual environmental, social, and governance (ESG) reports that outline their commitment to managing the environmental and social risks associated with their business models, along with their plans to achieve "sustainable growth" within their operations, supply chains, and business relationships. However, few are willing to acknowledge that boundless corporate expansion is not sustainable, but rather fuels climate change, often resulting in adverse impacts on families, communities, and economies. Based on the planetary boundaries, the continued expansion of industrial activity and economic development cannot be sustained, which restricts the ability of industries and markets to engage in unchecked growth responsibly. Further, by ignoring this restriction, companies contribute to the ongoing oppression of families, communities, and economies. Yet, most would agree that society needs economic development and industrial growth in some form to sustain, evolve, and continue prosperity. However, the point at which the benefits to societal prosperity derived from corporate expansion and economic development peak has less consensus. This chapter discusses the new prosperity model through the lens of private industry considerations. Impacts on the private sector are reviewed through a discussion on the purpose of a corporation, the relationship between ESG and financial performance, and the cost of boundless growth.

The Purpose of a Corporation

A good starting point for how best to address this challenge is to understand how the private sector and corporations evaluate success. Today, most companies measure success based on financial performance determined by market growth, or by an increase in financial assets obtained through corporate expansion within an industry. Behind this belief is the classical *shareholder primacy* model. First introduced in 1919 in the Michigan Supreme Court case *Dodge v. Ford Motor Co.*, the principle of shareholder primacy has been considered a tenet of the fiduciary duty owed by corporate directors for over a century, encouraging short-term profit-seeking behavior and often resulting in significant adverse impacts on society and the planet (Lee, 2023). The model, based on the idea that a corporation's sole responsibility is to make as much money as possible for its shareholders by protecting shareholder value at all costs, was lauded by economist Milton Friedman, who vehemently promoted free markets and rejected Keynesianism, believing that policymakers should control the amount of money in circulation. In his famous 1970 op-ed in *New York Times Magazine*, Friedman proclaimed that "there is one and only one social responsibility of business—to use its resources and engage in activities designed to increase its profits."

The decades-long debate around the purpose of a corporation is an interesting one. On one side, Friedman argues that any purpose assigned to a corporation outside that of profit maximization relies upon political mechanisms, not market mechanisms, for the allocation of scarce resources at the detriment of free markets and firm value. Interestingly, more than 50 years after Friedman penned his op-ed blasting the "schizophrenic" and "short-sighted" proponents of corporate social responsibility, a growing body of evidence points to strong ESG performance as a good indicator of long-term firm value. Further, on April 19, 2019, the Business Roundtable, a group of more than 200 CEOs leading the top companies across the United States, published a new statement on the purpose of a corporation that moved away from shareholder primacy toward the ***multi-stakeholder theory***, outlining that the purpose of a corporation is to benefit all stakeholders. The announcement came following decades of roundtable statements on corporate governance aligned with shareholder primacy as motivation for corporations to increase stock prices in all decisions, often to the detriment of other stakeholders in the name of profit maximization.

In contrast to the shareholder primacy model, the multistakeholder theory considers the impacts on all stakeholders, including employees, customers, suppliers, and impacted community members, not just shareholders, when making decisions. For example, in addition to protecting shareholder value, improving nonfinancial performance, such as reducing employee turnover, increasing internal promotion rates, or improving employee participation in resource networks, is valued by a company's employees. In addition, other stakeholders, such as a company's customers, are likely to value improved performance around managing customer complaints, data privacy, and improved product health, safety, and innovation. Further, suppliers and community members are often concerned with how a company mitigates the risk of forced labor within its value chain, the associated impacts on community development stemming from its operations, and how well it manages waste, reliance on natural resources, and its commitment to responsible sourcing.

The multistakeholder theory promotes the idea that corporations should track and measure success defined by increasing value for all stakeholders. Tracking corporate performance of this kind can be divided into two categories: (1) those that measure success tied to shareholder concerns, known as financial performance, and (2) those that measure success based on the needs of all other stakeholders, known as nonfinancial performance. Figure 14.1 illustrates the two main types of key performance indicators (KPIs)—financial and nonfinancial—and the main components of each type. Revenue, profit, and capital are financial performance indicators that firms commonly use to measure and analyze corporate success. Nonfinancial performance indicators are divided into three buckets—environmental, social, and governance—which comprise ESG performance. It could be argued that financial performance is to the shareholder primacy model what nonfinancial performance is to the multistakeholder model.

Historically, corporations have relied on KPIs to track and report on financial performance, which is often measured by currency indicators, such as an increase in financial assets or market valuation. On the other hand, nonfinancial performance indicators are just starting to gain traction in mainstream markets and are usually measured by societal and environmental indicators, such as a decrease in carbon emissions or an increase in the diversity dimensions of a workforce. Figure 14.2 provides examples of commonly used financial and nonfinancial performance indicators in Corporate America. Financial performance measures are well known,

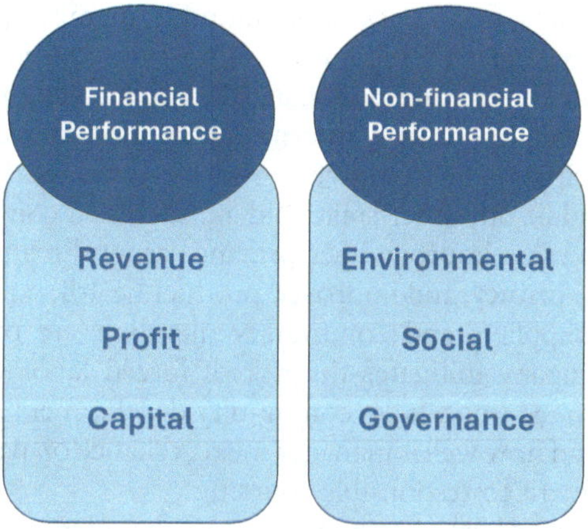

Fig. 14.1 Financial and nonfinancial performance

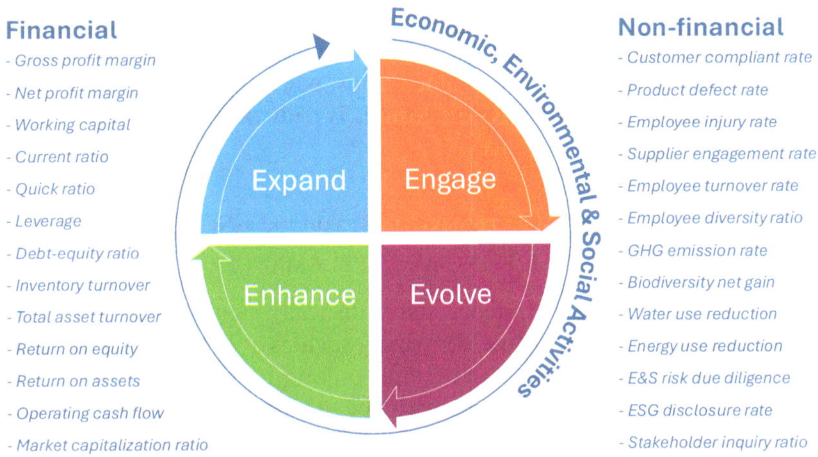

Fig. 14.2 Corporate key performance indicators

widely accepted, and frequently incorporated into corporate strategic planning. Conversely, nonfinancial performance measures are less common, lack standardization, and are used sporadically by firms today, often with nonbinding terms. While nonfinancial metrics are gaining traction with CEOs and business leaders, it is rare for a company to prioritize nonfinancial KPIs over financial ones, even if the long-term viability of a company depends on economic growth that is sustainable and responsible.

In 2021, economist Colin Mayer published a new framework to address the future of the corporation around three sets of issues: increasing environmental degradation, social inequality, and corporate corruption. In it, Mayer argues that the common debate between economists and sociologists over shareholder primacy versus multistakeholder theories misses the point, as “the efficient functioning of markets requires that the business of business is not just business and the governance of business is the business of government” (Mayer, 2021, p. 889). Rather, we must get back to the purpose of an organization, which requires an understanding of why a business was created and what its goal is, which Mayer believes should be derived from a system that produces profitable solutions to societal problems while also not profiting from creating societal problems. Yet, with more and more studies demonstrating that benefits are derived from both financial and nonfinancial performance, investors are increasingly asking questions about the ESG performance of a company and placing higher value on those companies actively engaged in both managing and disclosing ESG risk throughout their value chains, which has led to many wondering if the demand for ESG disclosures is truly driven by market value or, rather, by guilty consciences.

The Relationship Between ESG and Financial Performance

Today, multiple studies evidence the statistically significant relationship between ESG and financial performance, such as a 2019 study that analyzed the real estate sector as an example of long-term value in the Group of Seven (G7), or the countries with the strongest economies, namely Canada, France, Germany, Italy, Japan, the United Kingdom, and the United States, over a five-year period and found a positive relationship between ESG disclosure and the firm's return on assets and return on capital (Almeyda & Darmansya, 2019). In addition, another group of researchers reviewed the "100 best corporate citizens" in the United States as declared by 3BL Media from 2009 to 2018 and found that a commitment to environmental sustainability, consistent socially responsible conduct, and rationalized governance mechanisms positively correlate with market value and are good predictors of future financial performance in the market (Qureshi et al., 2021). The study found the correlation between environmental indicators and financial performance to be the strongest among all three nonfinancial performance measures, which prompts the question: is environmental sustainability the foundation of a healthy economy?

In 2022, researchers reviewed the financial data and ESG scores of 150 publicly traded companies listed in the Standard and Poor's 500 index during the years 2017 through 2020 and found that companies with superior ESG performance perform better financially and are valued higher in the market compared to their industry peers. Specifically, the study concluded that a company's ESG rating impacts both return on capital employed as a proxy for financial performance and Tobin's Q as a proxy for market valuation (Ademi & Klungseth, 2022). Another study in 2023 reviewed similar data of 3332 companies over a ten-year period, between the years 2011 and 2020, and found that ESG performance positively correlates with corporate performance for large-scale companies. However, the same correlation was not present for small-scale companies (Chen et al., 2023). The findings beg the question: does the value of ESG for companies depend on systemic scaling of ESG measures across industries, markets, and economies?

While studies show the importance of measuring success based on nonfinancial indicators, financial performance and economic growth are rarely tied to the principles of environmental sustainability and social responsibility, which are often retrofitted to an already-baked corporate strategy. Therefore, a new model for measuring if, when, or how the expansion of a company within an industry or market is deemed responsible and sustainable is desperately needed, one that considers both the micro and macro impacts of corporate expansion and economic development on families, communities, and economies, and requires environmental and social impacts stemming from growth to be evaluated upfront during strategy development, not as a wishful afterthought. We need a new model for measuring how economic development contributes to the prosperity of healthy families, communities, and society, while also accounting for the adverse environmental and social impacts associated with corporate expansion, consumerism, and the resulting climate change. We need a prosperity model that assesses not only if the growth of a business, industry, or market truly is economical, but also if it is truly sustainable and responsible.

A practical application of a new prosperity model is during the annual review and strategic planning process at the end of each fiscal year when a company assesses its performance and results for the previous year and sets its goals and targets for the upcoming one. Under a new prosperity model, a plan for corporate expansion would need to meet a four-part test to determine if the growth of assets is economical, responsible, and sustainable and has transparently assessed its long-term impacts. Absent confirmation that growth in an industry or market meets an immediate societal need economically, manages capital inputs and outputs responsibly, and establishes a value chain sustainably, a corporate plan for "sustainable growth" remains an oxymoron due to the inconvenient truth that blind, unchecked growth is not sustainable nor responsible and, therefore, not viable within a new theory of prosperity. As such, in these situations, a company should focus on nonfinancial performance indicators for the upcoming year, which still maintain and often increase the valuation and long-term viability of a company.

The Cost of Boundless Growth

Currently, most companies stress boundless market growth goals each year because traditional approaches to measure firm value emphasize the need for increased physical inputs to accelerate the accumulation of physical assets, which is how success has historically been measured. However, unchecked corporate expansion as the primary indicator of success is what is driving climate change. Yet, while some sociologists may advocate for abandoning capitalism altogether, the idea that economic growth only harms the planet and society is hard to get behind. Therefore, a better approach to measuring prosperity is to balance financial performance with nonfinancial performance by advocating for a shift in how we identify, measure, and protect firm value. In addition to financial metrics, success criteria should include measures such as employee well-being, customer complaint remediation, nature-regenerative projects, community engagement, and other nonfinancial indicators that evidence if a company's operations are sustainable and responsible, not only profitable.

Indeed, balance is key to sustainable, responsible growth. By understanding the relationship between managing the environmental and social risks associated with business operations and protecting firm value, firms can make better-informed, risk-based decisions that not only address the needs and concerns of all their stakeholders, but also protect the organization's long-term viability and contribute to its overall corporate performance. Figure 14.3 illustrates a New Prosperity Model, which is designed to assess not only if corporate growth is economical, but also if it is truly sustainable and responsible. The model balances corporate engagement, evolution, enhancement, and expansion with the needs of and impacts on society, the planet, and the economy, while also considering the reliance on social adaptation, financial capital extraction, technological adaptation, and natural capital extraction, all of which are crucial to realizing long-term prosperity.

The idea that a company should not prioritize market growth as a mandatory yearly goal or as its primary indicator of performance may appear crippling to a corporate board's long-term vision or counterintuitive to tenured business leaders. However, understanding the relationship between corporate expansion, climate change, and prosperity will sharpen one's view. For example, due to the closely woven relationship between ESG performance and financial performance, limiting market growth deemed to be irresponsible for a 12-month period to improve operations

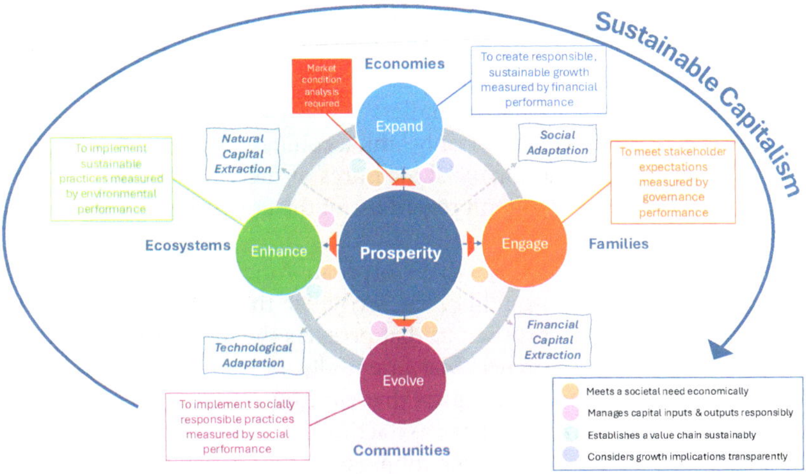

Fig. 14.3 A new prosperity model

or business relationships may protect the ability to expand responsibly in another year. Further, in circumstances where growth is not viable, a company can look to nonfinancial KPIs as its measure of success, such as increasing operational efficiency, delivering excellence in customer service, understanding evolving customer needs, or promoting innovation to address societal challenges. At the core of this dilemma is the deeply embedded myth that we are all conditioned to accept: that firm value is solely dependent upon and derived from ongoing, boundless market expansion.

With numerous studies evidencing that firms derive value from activities other than or in addition to financial performance, what is preventing companies from incorporating ESG into their strategic planning? Whatever the motivation may be, the stronghold on financial performance as the sole indicator of a firm's success persists, along with economic growth measured primarily by financial statements, which both support and are supported by traditional approaches to extractive capitalism. Within the historical growth model, firms prioritize market expansion as their number one overarching goal, which is believed to contribute to the health of economies above all else. Take, for example, the stock market of the 1950s compared to the value of financial assets today. According to FRED, the value of financial assets in 1950 was about equal to the GDP, or the movement of income and spending in the economy. Today, financial assets are valued at about five times the GDP. Yet, the economic measure of success continues to be tied to the boundless growth of these assets. Today's financial performance goals not only limit a company's ability to recognize and take advantage of nonfinancial measures of success, but also promote unchecked corporate expansion regardless of the environmental and social consequences.

While most Americans agree that Corporate America is driven by shareholder value and the bottom line, few recognize the societal conditioning that we are all regularly exposed to that protects the model of shareholder primacy. In 2023, researcher Marjorie Kelly published her manifesto, *Wealth Supremacy*, where she argues that wealth supremacy, or our collective bias that legitimizes limitless extraction for the wealthy, and capital bias, or our acceptance of worker subjugation within capital enfranchisement of votes by corporations, are the root causes of many of society's systemic problems. Kelly points out that we have all been conditioned to accept a rising stock market as the measure of economic success, while ignoring (or perhaps not making the connection to) the mass layoffs or sharp increases in natural resource extraction that are often what feed the bottom line to increase profits and drive rising stock prices (2023).

Take, for example, Exxon Mobil, which, according to Forbes, is one of the world's largest publicly traded oil and gas companies, with a market capitalization of $520 billion. Forbes often publishes guidance on the top oil and gas companies to invest in, outlining how increased crude oil prices incentivize companies to increase exploration and production activities (Forbes, 2024). So, in 2022, when Exxon Mobil's stock price increased by 80%, it was mostly due to two factors: (1) high crude oil prices and (2) the shareholder primacy model that normalizes maximizing gains to shareholders, even at the cost of uncontrollable wildfires, rising sea levels, and extreme tropical storms. At the core of extractive capitalism is the idea that our economy exists to extract limitless resources with the goal of growing and protecting financial capital at all costs, including the destruction of ecosystems and their vital services.

Exxon is one example of how the power and influence of one global conglomerate can drastically shift a market quickly and sharply. Therefore, does a new model work if not everyone buys in? While a new prosperity model may appear to work in theory, what is to prevent a large global company, or a small coordinated group of companies, from upending the model? In other words, if 95% of companies adopt a new measure of prosperity, what is to stop the other 5% from going rogue or continuing business as usual, and preventing society (and economies) from realizing the full potential of a new model of prosperity? The four boundaries—engage, evolve, enhance, and expand—are each associated with unique challenges and opportunities in the market, but what is the governance mechanism to ensure companies heed the risks appropriately and seize the benefits accordingly? This is where the role of public policy comes in.

References

Ademi, B., & Klungseth, N. J. (2022). Does it pay to deliver superior ESG performance? Evidence from US S&P 500 companies. *Journal of Global Responsibility, 13*(4), 421–449.

Almeyda, R., & Darmansya, A. (2019). The influence of environmental, social, and governance (ESG) disclosure on firm financial performance. *IPTEK Journal of Proceedings Series, 5*, 278–290.

Chen, S., et al. (2023). Environmental, social, and governance (ESG) performance and financial outcomes: Analyzing the impact of ESG on financial performance. *Journal of Environmental Management, 345*, 118829.

Forbes. (2024). 5 of the best oil and gas stocks to buy for high crude prices. *Forbes*. Retrieved May 16, 2024, from https://www.forbes.com/sites/investor-hub/article/best-oil-gas-stocks-amid-high-crude-oil-prices/?sh=7b4ca6af2b73

Kelly, M. (2023). *Wealth supremacy: How the extractive economy and the biased rules of capitalism drive today's crises*. Berrett-Koehler Publishers.

Lee, Y. S. (2023). Shareholder primacy as an untenable corporate norm. *Annals of Corporate Governance, 8*(1), 1–50.

Mayer, C. (2021). The future of the corporation and the economics of purpose. *Journal of Management Studies, 58*(3), 887–901.

Qureshi, M. A., et al. (2021). Do ESG endeavors assist firms in achieving superior financial performance? A case of 100 best corporate citizens. *SAGE Open, 11*(2), 21582440211021598.

United Nations. (2024). *Global Resources Outlook 2024: Bend the trend-Pathways to a liveable planet as resource use spikes.*

World Economic Forum. (2024). *The global risks report 2024 19th edition*. Cologny, Switzerland.

CHAPTER 15

Public Policy Implications

The role of public policy is to make decisions and take action to address societal issues using the tools, mechanisms, and interventions available to policymakers. If we determine that climate change is an issue negatively impacting societal prosperity, then we must also determine that it is the responsibility of public policy to address it. Most policymakers, regardless of party affiliation, campaign on promises to promote and protect the health of families, communities, and economies. Therefore, policymakers must address the climate crisis as a root cause of promoting and protecting prosperity. Yet, few are willing to acknowledge a changing climate as a root cause of many of the challenges and issues plaguing American families, communities, and economies, let alone act on climate policy. Some policymakers are even introducing or supporting new laws and regulations aimed at prohibiting private enterprise, particularly banks, from limiting investments in certain sectors associated with adverse environmental and social impacts based on the perception that restricting access to capital for carbon-intensive industries is a form of discrimination. This chapter discusses the new prosperity model through the lens of public policy implications. Impacts to the public sector are discussed through the role of public policy, rational sustainability, and the role of economics in public policy.

S. Haley Knowles, *Diminished Prosperity*,
https://doi.org/10.1007/978-3-032-07068-5_15

The Role of Public Policy

On January 14, 2021, following nearly two years of public announcements by several banks of their commitment to reach net-zero emissions by the year 2050, the Office of the Comptroller of the Currency (OCC) published its final *Fair Access to Financial Services* Rule, which prohibited covered banks from denying a financial product or service to a customer that would prevent the customer from "entering or competing in a market or business segment." The rule was largely in response to banks' new voluntary ESG disclosures that outlined how they conduct due diligence on high-ESG-risk sectors, mostly carbon-intensive industries or those associated with human rights abuses, and restrict the financing of sectors deemed to be outside their risk appetite. The rule was also driven by the belief that by restricting investment in carbon-intensive industries, banks were "discriminating" against oil and gas companies, among others. While the final rule acknowledged the right of banks to provide or deny financial services to any current or prospective customer, it emphasized that banks must first "do their homework and be able to show their work…[based on] measurable, empirical, quantifiable data evaluated under the bank's established, impartial risk-management standards established in advance by the bank."

Shortly after the Biden Administration took office, the OCC "Fair Access" Rule was rescinded, but not before inspiring policymakers across the United States at all levels to introduce their own versions. As a result, a number of states have enacted, are enacting, or are considering enacting legislation to require financial institutions to provide commercial clients "fair access" to financial services, including Texas, Florida, Arizona, Georgia, Idaho, Indiana, Iowa, Kentucky, Louisiana, and South Dakota (Sullivan & Cromwell, 2024). By applying guidance designed for consumer protection from discrimination based on protected classes, policymakers are prohibiting banks from "discriminating" against oil and gas companies, firearm manufacturers, and other sectors often associated with adverse environmental and social impacts through the use of ESG assessments. However, it appears these laws may not be protecting the safety and soundness of financial services as they claim but, instead, may be driving down fair competition in markets and, as a result, decreasing options and increasing costs to taxpayers.

For example, a 2023 study analyzed the impacts of Texas Senate Bill 13 and Senate Bill 19, which were both enacted in 2021 to prohibit state municipalities from contracting with banks that restrict funding to oil and gas companies or firearms manufacturers. Study findings show that the new laws resulted in five of the largest underwriters, namely Goldman Sachs, Citigroup, JPMorgan Chase, Bank of America, and Fidelity, exiting the Texas bond market, resulting in a significant decrease in competition and an increase of up to $532 million in interest (Garrett & Ivanov, 2024). Further, the new laws require banks to annually attest to the State Attorney General that they comply with the law, while banning those banks that decline to amend their ESG practices from doing business in or with the state. As a result, another study conducted in 2024 by the Texas Association of Business Chambers of Commerce Foundation found that the same laws will cost the State of Texas approximately $668.7 million in lost economic activity; $180.7 million in decreased annual earnings; 3,034 fewer full-time, permanent jobs; and $37.1 million in lost tax revenue. Shockingly, "fair access" laws, those that seek to outright ban ESG as a financial or corporate practice that unjustly handicaps certain companies and industries purely for political and personal beliefs or reasons, appear to be hindering fair competition in certain financial markets.

In addition, diversity, equity, and inclusion (DEI) programs have come under attack in recent years. Following the June 2023 Supreme Court decision in *Students for Fair Admissions,* which effectively upended affirmative action programs across universities and corporations, companies started backtracking on their DEI efforts, or those meant to promote the recruiting, retaining, fair treatment, and full participation of employees from various backgrounds, abilities, and identities through policies and initiatives designed to help all workers feel welcomed and equipped to perform their jobs at a high level as an integral component of a company's success (Yoshino & Glasgow, 2024). As a result, while most companies perceive DEI efforts as part of a larger sustainability strategy and vital in addressing systemic inequities and issues that prevent or hinder the realization of prosperity for all, the increased politicization of sustainability issues across all pillars and domains is transforming the way the private and public sectors approach sustainability. Further, the politicization of sustainability initiatives brings with it the polarization of ESG concepts and ideals across all stakeholder groups, and often irrational decision-making.

In May 2024, *The New York Times* reported that Ron DeSantis, governor of Florida—the state where studies show that 2023 was the hottest year on record, with coastal waters rising to over 90 degrees Fahrenheit and Hurricane Ian, the 2022 hurricane that killed over 140 people and caused more than $109.5 billion in damages, is now the costliest hurricane in state history, signed into law a bill that not only eliminates state funding for energy conservation and renewable energy programs but also removes requirements that state agencies purchase fuel-efficient vehicles, among other things. The bill became law on July 1, 2024, during a time when insurance companies are refusing to provide adequate coverage to many Florida homeowners and, for those who are approved for coverage, charging some of the highest rates in the country, and only a year after the same governor rejected $346 million in federal funding to make Floridian homes more energy efficient (Davenport, 2024). Any reasonable person observing what is happening in Florida would be correct in asking: if residents living in high climate-risk regions are being adversely affected by extreme storms, increased temperatures, and the associated fallout of climate change, why then are policymakers enacting laws that appear to be sustaining, even fueling, climate change?

Rational Sustainability

The onslaught of anti-ESG legislation has swept across the United States, confusing many ESG practitioners and complicating well-intentioned sustainability programs. It appears policymakers are enacting laws against the best interests of their constituents, or even their own best interests (in the long term). The resulting polarization of what should be approached as evidence-based decision-making is muddying the waters for sustainability leaders in both private and public sectors. This dilemma is explored by Alex Edmans in his recent article *Rational Sustainability* (2024). In it, Edmans acknowledges that the goal of sustainability, which is to create long-term value, not slap ESG labels on companies or projects to make them more appealing, is something that everyone can get behind and, therefore, the issue is not the intended outcome but the approach to ESG. He advocates for a shift toward rational decision-making that addresses not only the benefits of sustainability but also the diminishing returns and trade-offs (Edmans, 2024). Yet, regardless of how we approach it, public policy has a lot of catching up to do.

In order for sustainable capitalism to thrive within a new prosperity model, policymakers must evolve economic policy to enable, incentivize, restrict, and monitor how private enterprise expands in the marketplace, meaning they must put forth policy solutions that not only enforce and guide the understanding of the preconditions needed for responsible, sustainable growth but also intervene when the strategic direction of private enterprise is at odds with current market conditions. However, these policy mechanisms are not new. In fact, most have been leveraged by policymakers for decades, even centuries. What requires evolving is not the approach to policy intervention but the focus and goal of the intervention. Setting irrational decision-making aside, modern policy interventions often seek to redistribute economic value amongst players in the market to address equity issues, fix prices to account for surpluses or deficits, or limit access to balance supply and demand. Yet, what most policy interventions achieve is limited to treating the symptoms of a broken system, not remediating the root causes of inequity or market dysfunction.

If we know that changes to our climate drive increased costs of food, housing, and healthcare, why are policies not aimed at mediating the causes of climate change? Further, if human-caused global warming is impeding our ability to freely choose our reproductive destinies, why are more people not demanding policymakers put forth effective climate policy? And, if the long-term impacts of unchecked global warming render undesired changes to market prices, why doesn't policy require companies to incorporate a changing climate into their strategic planning decisions? The role of public policy should always be to intervene when the prosperity of families, communities, and economies is at risk. The idea of free markets without government intervention was coined during a time when economists did not have to factor in the adverse impacts of a changing climate as an externality and, therefore, can no longer be upheld as a societal principle when it is rarely relevant to the conditions of a modern society.

Figure 15.1 outlines the role of public policy to address and solve societal problems within a new prosperity model, particularly the need for policymakers to incentivize, require, monitor, and oversee the design, execution, and measurement of strategic planning, market expansion, and corporate performance in Corporate America with a mix of preproduction, postproduction, and production-stage policy interventions. Along with balancing corporate engagement, evolution, enhancement, and expansion with the needs of and impacts to society, the planet, and the economy, the model highlights the important role public policy plays in balancing environmental, human, and economic sustainability.

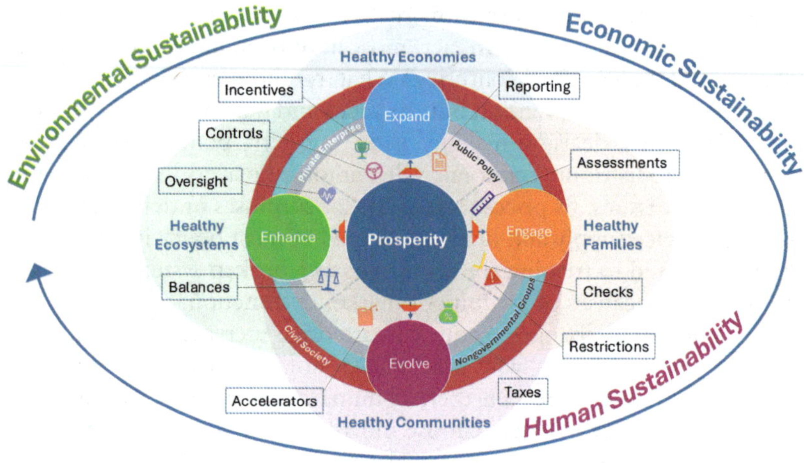

Fig. 15.1 Prosperity and the role of public policy

The Role of Economic Policy

While classical economic theory has some catching up to do in regard to today's changing climate, there is still much that is relevant to a modern society's quest for prosperity. For example, the *Nash equilibrium* demands that policy outcomes consider the resulting impacts to everyone, as the outcome that is determined to be best for all is also the outcome deemed best for each individual. When applying this concept to financial and non-financial performance measures within the new prosperity model, Nash's theory promotes the need for market preconditions to determine the strategic direction of a company. Without this analysis, it is easy (and likely) for both policymakers and corporations to make decisions based on what is best for their balance sheet and stock price instead of considering what is best for society, the environment, and the economy. Applying the *Nash equilibrium* to the new prosperity model resembles the modern debate between the shareholder primacy and the multistakeholder model. Perhaps the dilemma is less about acknowledging or understanding the impacts of climate change and more about determining whose impacts matter.

Most of what we know today about economic theory was developed during a time when women could not vote, own property, or even hold a job in most industries, let alone make a fair wage. This means that classical economic theory was developed by men at a time when they did not need to (or think to) consider the impacts to the labor market if worker productivity diminished regularly over a nine-month period, often capped by a short-term sharp decrease in productivity altogether. While public policy has caught up on gender equity in some areas, systemic inequities remain, rooted in a system that was designed hundreds of years ago by men, for men. How different would classical economic theory be if those who wrote it regularly experienced pregnancy, childbirth, and caring for a newborn baby? Further, what if those who designed food, health, housing, and climate policy today were mostly Black and Indigenous women, or those most impacted by a changing climate?

Digging into this idea further, Smith's *invisible hand* underscores who benefits from market forces, claiming that we all derive value when we each make rational decisions in our own self-interests. This theory provides another example of policy mechanisms that could benefit the new prosperity model, but only if we assume that all players in the market are rational, meaning that when they act in their own self-interests, they are also acting in the best interest of everyone. Yet, knowing that the majority of policymakers today are affluent white men, and thinking about Edmans's *rational sustainability* theory, we must consider the possibility that many policymakers, along with corporate boards, engage in irrational activity at times, or, in other words, make decisions against their own best interests in the long term by focusing on their own gains in the short term. Could it be that policymakers are engaging in irrational decision-making because they are focused on what benefits their own self-interests in the here and now?

Perhaps the answer can be found in Bataille's *excess of energy*, where a surplus in the market demands allocating, either productively or recklessly. In historical growth models, the approach and method to excess is governed by only one consideration: what best serves the individual. However, by applying this theory to a new prosperity model, excess energy is more likely to be spent productively due to the market preconditions that must be met before growth in a certain area is deemed viable. When growth is determined to not be productive, excess energy is channeled into other productive endeavors which contribute to both financial and nonfinancial performance and, ultimately, benefit all of society. Perhaps policymakers need reminding of their duty to represent their constituents. A 2023 *Harris poll* found that more than half of Americans, across race, ethnicity, age, gender, and political affiliation, believe that extractive capitalism is headed in the wrong direction and no longer benefits them or their families.

While many references to the historical work of sociologists and economists can be leveraged to demonstrate the importance of nonfinancial performance measures, the most compelling may be Nietzsche's *test of the eternal return*. In his theory, Nietzsche asks us to consider the livability of experiencing the same moments in our life over and over again and, in doing so, conclude whether our choices and the resulting consequences are worth experiencing in perpetuity. The concept, while uncomfortable to some, shines a light on the human condition as a consequence of our collective choices and implicates our actions today as consequences for younger generations tomorrow. Therefore, if applied to policymakers and corporate boardrooms, the *test of the eternal return* forces both to consider the long-term consequences of every decision they make, which also forces them to consider the impacts of a changing climate in the design, execution, and oversight of their annual strategic plan, priorities, and goals.

The only way a new prosperity model works is if public policy does its part to implement the governance mechanisms needed to ensure all market players are playing fairly before and during the match, not simply intervene after the game has been played in an attempt to balance the score. Here, the role of public policy depends on early production-stage policy intervention, which is often rare. Policymakers must be willing to intervene before corporate expansion wreaks havoc, intentionally or unintentionally, on an ecosystem, a community, or an economy. While oversight and enforcement actions after the fact are needed to correct when irresponsible or unsustainable growth has taken place, preproduction and postproduction policy interventions cannot be solely relied upon for oversight of a new prosperity model. Options regarding form, format, and delivery are vast and wide-ranging. Timing, however, is less flexible.

While progress on climate action remains slow, there has been some movement on climate in the public policy arena. For example, on March 6, 2024, the Securities and Exchange Commission issued a final rule that requires registrants to provide climate disclosures in their annual financial statements, including effects of severe weather events and other natural conditions, carbon offsets and renewable energy certificates, and financial impacts from severe weather events and climate-related targets or transition plans. While the final rule is tied in federal litigation, the regulation comes months after the State of California passed its own law with similar requirements. A handful of other states, including New York, New Jersey, and Colorado, have followed suit. In addition, the year prior, on July 31, 2023, the European Union adopted the Corporate Sustainability Reporting Directive, which requires more than 50,000 companies doing business in Europe, including tens of thousands of companies based in the United States, to annually disclose the environmental and social impacts within their business model, strategy, policies, and due diligence, including climate risk management, respect for human rights, oversight of anti-corruption and bribery issues, and workforce diversity dimensions. While additional rules around the managing of and reporting on environmental and social risks within value chains are needed, the adoption of all three laws affirms the importance of nonfinancial performance measures on a firm's annual disclosures and provides hope that the evolution toward a new prosperity model may be possible.

With all the recent studies coming out of academia and the reports being released by governmental agencies, we have plenty of evidence that climate change is positively correlated with factors contributing to significant financial strain to American households such as increased destruction to property, higher expenditures on groceries, and added healthcare costs from heat-related medical needs. With increased knowledge of the positive relationship between global warming and costs of living, the implications of a changing climate to society are clear. Climate change directly impacts the prosperity of families, communities, and economies, impeding livelihoods through increased costs of living, which exacerbate parenting costs, influence family planning decisions, and prevent households, communities, and economies from healthy, sustainable growth and resiliency. In the era of climate change, traditional approaches to economic models fall short. Therefore, we need a new prosperity model for the private sector, one that is grounded in sustainable, responsible growth, while also responding to market conditions and the needs of society, with proper governance and public policy oversight.

Over the last decade, the idea (or hope) that technological advances will be enough to address climate change before catastrophic events occur has gained popularity among many optimistic sociologists, economists, policymakers, and everyday inhabitants of Planet Earth. While the idea of technology saving us from ecological disaster is compelling, it cannot account for the fact that climate change is a multidimensional, multifaceted, complex, and systemic problem with drivers deeply embedded in all areas of society. We need to rethink technology as the silver bullet for mitigating and managing climate risk and start having the difficult, uncomfortable conversations about how our lifestyle choices and purchasing decisions uphold unchecked corporate expansion and, ultimately, are the main driver of global warming. Unfortunately, there is no silver bullet for solving the climate crisis, and technology alone will not save us. The only effective pathway for addressing the climate crisis and its associated adverse impacts is through deep, systemic change that requires a new measurement for corporate expansion and success. A good place to start is with a new prosperity model that assesses not only if the growth of a business, industry, or market is economical, but also if it is truly sustainable and responsible.

References

Davenport, C. (2024). Deleting climate change from Florida policy. *The New York Times.* Retrieved May 15, 2024, from https://www.nytimes.com/2024/05/15/climate/desantis-climate-change-florida.html

Edmans, A. (2024, January 20). *Rational sustainability.* London Business School.

Garrett, D., & Ivanov, I. (2024). *Gas, guns, and governments: Financial costs of anti-ESG policies.* Jacobs Levy Equity Management Center for Quantitative Financial Research Paper.

Sullivan & Cromwell LLP. (2024). *States require "Fair Access" to financial services.* Retrieved May 1, 2024, from https://www.sullcrom.com/SullivanCromwell/_Assets/PDFs/Memos/States-Require-Fair-Access-Financial-Services.pdf

Yoshino, K., & Glasgow, D. (2024). DEI is under attack. Here's how companies can mitigate the legal risks. *Harvard Business Review*. Retrieved April 4, 2024, from https://hbr.org

CHAPTER 16

A History of Societal Evolution

I spent the first 20 years of my life in South Dakota on occupied Očhéthi Šakówiŋ, better known as the Great Sioux Nation, and Yankton land. In 1989, before I began formal education and two decades after the federal recognition of Columbus Day, the South Dakota State Legislature unanimously voted to replace Columbus Day with Native American Day, becoming the first state in the United States to do so. Yet, it would take another two decades, when I took my first job after undergrad with the Social Security Administration and moved away from South Dakota, for me to understand the impact of this change on me. Sitting at my desk one morning in early October, I remember seeing Columbus Day on the federal calendar and feeling both shock and anger that, outside of South Dakota, people celebrated the day differently than how I was raised.

Growing up, I have many memories of learning about the bravery and selflessness of great Chiefs and of the beauty of Indigenous lifeways each year in school on Native American Day. I remember one year building a replica of the great city of Tenochtitlan out of sugar cubes. Another year, I baked bread with my classmates alongside a Lakota woman who told us how her ancestors handed down the sacred recipe and accompanying story of family and friendship from generation to generation. As I got older and developed new friendships, I learned of tribal ceremonies and fascinating customs from my Lakota friends. My friend Nikki was the first to tell me a story of the Giveaway Ceremony, one of the many stories that had been

S. Haley Knowles, *Diminished Prosperity*,
https://doi.org/10.1007/978-3-032-07068-5_16

shared with her by her Lakota elders of the great leaders who lived before her and the beautiful horses and blankets they gave away to their community at their giveaways, known as the one true measure of success and prosperity for her, her people, and their way of life.

When settlers first hear of the revered Giveaway Ceremony, it is often confusing to them due to the conditioning of the settler state that begins with children from an early age, rendering it difficult to reconcile giving away wealth to others as a symbol of success while also sustaining status for oneself. Yet, how different would our understanding of prosperity be if the historical achievements shared with children in school were not focused on settler colonialism? Can you imagine what our society would look like if we based success not on degrees and titles but on what we give back to our community? Or if the stories of great leaders that we told were not about their conquests of other cultures, lands, and people but about what they gave away to others as a symbol of their prosperity? How differently would children envision their career paths or college majors if the probing questions from uncles and grandparents at family gatherings centered around a measure of success dependent upon giving? What would be the societal impact of the generational shift that would occur from embedding the teachings of Indigenous lifeways in all early childhood education?

True prosperity depends upon the realization of a few core principles. The first being full bodily autonomy, or the right to freely choose to have a child, to not have a child, and to raise a child in a safe and healthy environment. The impacts of climate change are vast, intense, and crippling, from rapid urbanization and homelessness to heat-related illnesses and soaring medical bills. However, the consequences of a changing climate are not shared equally. Women have historically experienced oppression at a higher rate than other demographics, particularly within communities of color on the frontlines of climate change, along with having their unique healthcare needs ignored when compared to the population at large. Further, with the adverse impacts of climate change compounding throughout communities, more families are experiencing increased costs of living, often resulting in family sizes decreasing, and not always voluntarily. Thus, the reclaiming of bodily autonomy through the lived experiences of women, particularly women of color, is essential for reproductive justice and prosperity to be realized. A reclaiming that is dependent not only on the right to full bodily autonomy and to raise a healthy child in a safe and healthy home, but also on the ability to determine when and if those rights have been fulfilled based on one's lived experience. For what is a better measure of a society's prosperity than the ability for all members to freely choose if, when, or how to have a child?

Second, prosperity must be underscored by a measure of progress that considers the lasting impacts on everyone, and from the vantage point of those most adversely impacted. However, today's corporate expansion, as a continuation of settler colonialism, exists to extract natural resources and worker productivity to the maximum in an effort to grow and protect shareholder value. And, with women of color being most impacted by climate change, modern corporate feminist empowerment campaigns aimed at liberating women are actually contributing to climate change and, ultimately, the oppression of low-income communities of color, especially nonwhite women. Therefore, a reconstruction of feminism is urgently needed, one that reimagines what prosperity means and takes a stand against white feminism's marriage to white supremacy, capitalism, and colonialism. We need to call out the relationship between feminism, often lauded as the path to equality in corporate success models, and climate change, and its collective understanding of prosperity through the shareholder primacy model. We need to decouple modern white feminism from consumerism and corporate colonialism and remediate white feminism's understanding of bodily autonomy, rooted in racist and classist policies. For if the modern, mainstream campaigns that claim to promote, create, and protect the prosperity of all, regardless of gender, race, or class, are actually contributing to the oppression of many in the name of progress, what hope do the disenfranchised have in realizing prosperity within a capitalist society?

And, finally, the measure of prosperity must go beyond economic value to include considerations of nonfinancial value based on impacts to the environment and society at large, not just those with vested capital. Classical economic theory was written at a time when climate change did not need to be considered as an externality. The shareholder primacy model was designed before knowledge of the adverse environmental and social impacts associated with business models was widespread, let alone well-researched and understood. The traditional corporate governance norms of today do not contribute to, protect, or promote the prosperity of all families, communities, and economies; they exist to serve only those with corporate voting rights. As a result, modern society's measure of success through maximizing shareholder value only contributes to the prosperity of the elite. Therefore, we need a new model of prosperity for measuring how economic development contributes to the prosperity of all families, communities, and societies, while also accounting for the adverse environmental and social impacts associated with corporate growth, consumerism, and the resulting changes to our climate. A new model of prosperity must be designed to assess not only if corporate growth is economical, but also if it is truly sustainable and responsible by balancing corporate engagement, evolution, enhancement, and expansion endeavors with the needs of and impacts on all of society, the planet, and the economy.

At first glance, a new prosperity model may seem like a sharp departure from everything we know about economics, sociology, political science, or progress. However, as a society, we have lived through many revolutions, all of which, at one time or another, seemed radical, impractical, or downright unattainable to many. For example, prior to the agricultural revolution, often referred to as the wide-scale societal transition from a hunting-and-gathering lifestyle to settlements built around farming, humans were a nomadic society that followed weather patterns and herds of animals to sustain our survival (Chu et al., 2022). The agricultural revolution brought with it the ability for people to grow food and establish homesteads, which led to cooperation among families, communities, and eventually civilizations. Towns and villages were built, which eventually led to schools, hospitals, markets, social gatherings, and formal political processes. Our foundation as a social, cooperative society can be traced back to humans' ability to domesticate animals and crops, and the markets, economies, and governments that arose from it.

Or consider the industrial revolution(s), which is remembered as the transformative societal transition from a communal barter and trade economy toward mercantilism supported by large-scale, stable manufacturing processes, where communities evolved from making goods by hand to machine production, resulting in great increases in labor efficiency and output (Clark et al., 2008). The industrial revolution brought with it the ability to produce merchandise much faster and cheaper, which expanded the opportunity for households to own consumer goods, increase comfort, and embrace convenience. As a result, industry expanded, jobs exploded, and earnings swelled, leading to an improved standard of living across all classes. In many ways, the industrial revolution changed how countries view the benefit and power of imports and exports, which were leveraged to solidify certain countries as global economic powers. The industrial revolution also sparked a curiosity in innovation, advancement, and the ability to improve societal conditions through invention and manufacturing.

Another example can be found in the more recent digital revolution, where society experienced a rapid shift from the traditional industries established during the industrial revolution to an economy centered around digital advancement and information technology. The digital revolution significantly changed the way information is processed, transmitted, shared, and stored, which also provided a foundation for Internet communications as the driving force of social evolution and opened doors to technological experimentation and innovation across all fields, including medicine, defense, and education. As a result, our world became more connected, with the ability to share ideas, experiences, and solutions quickly and conveniently across all geographical regions (Helbing & Helbing, 2015). All of these revolutions completely upended the way people lived, worked, and raised their families, building upon newfound knowledge that there is a better way to live. And still, we continue to evolve, to get better, and to change our way of life when we discover that a better way is not only possible but also needed for our long-term survival.

We are now in the midst of a sustainability revolution, driven by studies, reports, scientific findings, innovation, and the knowledge that extractive capitalism no longer meets our needs (and maybe it never did). Once complete, the sustainability revolution will have transformed everything we know about societal progress, economic growth, and their success criteria, indicators, and measurements. The sustainability revolution will reorganize our economy around low-carbon energy sources, realign social and economic prosperity with the health of families, communities, and the planet, and evolve boundless extractive methods of production toward ones built around sustainable, responsible growth. At the core of the sustainability revolution is the ability of societies to provide for their long-term survival. What may seem radical, impractical, or unattainable today is the start of a societal revolution that will transform, yet again, the way humans live, work, reproduce, and measure our collective prosperity.

The new model is based on the tenets of the reproductive justice framework, which provides a rare view into a world designed around and for freedom. The model exemplifies the role of corporations, policymakers, and all impacted stakeholders by providing a pathway for corporate expansion that does not infringe on individual bodily autonomy, or the ability of individuals to freely choose to have a child, to not have a child, and to parent a child in a safe and healthy environment. Under a new prosperity model, the realization of both individual fertility desires and collective economic prosperity may be possible, which together can create the highest level of societal prosperity for all.

Yet, while I offer a new model for measuring prosperity, one I believe is better fit and able to promote, protect, and understand the longevity of markets, economies, communities, and ecosystems, just as technology is not a silver bullet for solving climate change, a new model of prosperity cannot and will not address or remediate all the problems and challenges associated with modern extractive capitalism. Systemic issues such as structural racism, classism, xenophobia, corporate greed, state corruption, misogyny, and a plethora of others are deeply rooted within formal and informal institutions within our society, contributing, at varying degrees and intensity, to the dysfunction and oppression of modern extractive capitalism. A new prosperity model seeks to prevent the irresponsible and unsustainable mode of corporate expansion fueling climate change, but the model is not designed to address all the root causes of poverty, inequity, or corporate greed, or even a majority of them. These systemic issues are complex, vast, diverse, and evolving, and each requires a series of related, multifaceted, but separate solutions designed by and for those most impacted by them.

Rather, a new prosperity model provides a view into how abating climate change, by remediating the highest contributing factor to human-caused global warming, interacts with other societal challenges; a view that could be used to better understand the correlation between extractive capitalism and food, housing, and healthcare insecurity, along with other societal, environmental, and economic challenges. As discussed, climate change did not create malnutrition, vagrancy, or medical debt, but it does increase the number of people experiencing each and add complexity to the experiences of those already living with them. However, additional research is needed to explore the effectiveness of the model in both the private and public sectors, along with if or how the model interacts with other root causes of increased food, housing, and healthcare costs or other issues facing society today, and the mode, metric, and fitness of various quantitative and qualitative measures for each market condition and their associated outcomes at both micro and macro levels.

Indeed, there are many possible extensions of this research, but the criticality of its implications demands that too much energy not be spent on where to start but rather on how quickly we can begin, for the tightly woven, deeply embedded relationship between climate change, reproductive justice, and economic prosperity can no longer be ignored. Further, while through this research I aim to make a valuable contribution to the growing body of research on environmental, social, and economic policy and join the emerging dialogue around corporate social responsibility and sustainability strategy and the proper measure of economic prosperity, I also hope to uplift the reproductive justice framework and take a step toward claiming accountability for my role in settler colonialism, and by doing so, encourage other white settler women to examine and acknowledge their role within corporate colonialism as a communal step toward decolonization, disrupting white feminism, and progressing toward a new model of shared prosperity for all.

References

Chu, A. C., et al. (2022). Agricultural revolution and industrialization. *Journal of Development Economics, 158*, 102887.

Clark, G., O'Rourke, K. H., & Taylor, A. M. (2008). Made in America? The new world, the old, and the industrial revolution. *American Economic Review, 98*(2), 523–528.

Helbing, D., & Helbing, D. (2015). *What the digital revolution means for us* (pp. 177–187). Springer International Publishing.

References

Abbas Khan, K., et al. (2019). Natural disasters and economic losses: Controlling external migration, energy and environmental resources, water demand, and financial development for global prosperity. *Environmental Science and Pollution Research, 26*, 14287–14299.

Ademi, B., & Klungseth, N. J. (2022). Does it pay to deliver superior ESG performance? Evidence from US S&P 500 companies. *Journal of Global Responsibility, 13*(4), 421–449.

Almeyda, R., & Darmansya, A. (2019). The influence of environmental, social, and governance (ESG) disclosure on firm financial performance. *IPTEK Journal of Proceedings Series, 5*, 278–290.

Alston, M. (2013). Introducing gender and climate change: Research, policy, and action. In *Research, action, and policy: Addressing the gendered impacts of climate change* (pp. 3–14). Springer.

Alston, M., & Whittenbury, K. (Eds.). (2012). *Research, action, and policy: Addressing the gendered impacts of climate change* (pp. 71–112). Springer Science and Business Media.

Anastario, M., Shehab, N., & Lawry, L. (2009). Increased gender-based violence among women internally displaced in Mississippi 2 years post–Hurricane Katrina. *Disaster Medicine and Public Health Preparedness, 3*(1), 18–26.

Arora-Jonsson, S. (2011). Virtue and vulnerability: Discourses on women, gender, and climate change. *Global Environmental Change, 21*(2), 744–751.

Asaduzzaman, M. (2015). *Livelihood vulnerability of women in the context of climate change impacts: Insights from coastal Bangladesh*. Unpublished PhD thesis, The University of Newcastle.

S. Haley Knowles, *Diminished Prosperity*,
https://doi.org/10.1007/978-3-032-07068-5

Awiti, A. O. (2022). Climate change and gender in Africa: A review of impact and gender-responsive solutions. *Frontiers in Climate, 4*, 895950.

Barber, A. (2021). *Consumed: The need for collective change; colonialism, climate change & consumerism*. Hachette UK.

Bardwell-Jones, C. T., & McLaren, M. A. (2020). Introduction to Indigenizing and decolonizing feminist philosophy. *Hypatia, 35*(1), 2–17.

Barkemeyer, R., & Figge, F. (2012). Fordlândia: Corporate citizenship or corporate colonialism. *Corporate Social Responsibility and Environmental Management, 19*(2), 69–78.

Bataille, G. (1949). *The accursed share: An essay on general economy*. La Part Maudite.

Baten, J., & de Pleijt, A. M. (2022). Female autonomy generated successful long-term human capital development: Evidence from 16th to 19th century Europe. *World Development, 158*, 105999.

Bear, S. Y. (1996). *Standing in the light: A Lakota way of seeing*. U of Nebraska Press.

Beck, K. (2021). *White feminism: From the suffragettes to influencers and who they leave behind*. Simon and Schuster.

Becker, G. S. (1960). An economic analysis of fertility. In *Demographic and economic change in developed countries* (pp. 209–240). Columbia University Press.

Becker, G. S. (1981). *A treatise on the family*. Harvard University Press.

Becker, G. S. (1992). Fertility and the economy. *Journal of Population Economics, 5*(3), 185–201.

Berliner, J. S. (1999). *The economics of the good society: The variety of economic arrangements*. Blackwell Publishers.

Blaut, J. M. (1989). Colonialism and the Rise of Capitalism. *Science & Society, 53*(3), 260–296.

Bloem, M. W., et al. (2010). Castel Gandolfo Workshop: An introduction to the impact of climate change, the economic crisis, and the increase in the food prices on malnutrition. *The Journal of Nutrition, 140*(1), 132S–135S.

Bourdieu, P. (1998). *Practical reason: On the theory of action*. Stanford University Press.

Bourdieu, P. (2018). Cultural reproduction and social reproduction. In *Knowledge, education, and cultural change* (pp. 71–112). Routledge.

Bourdieu, P., et al. (1977). *Education, society, and culture* (pp. 15–29) (R. Nice, Trans.). SAGE Publications.

Bowen, A., & Hepburn, C. J. (2012). *Prosperity with growth: Economic growth, climate change and environmental limits*. Working paper, (93).

Bradbear, C., & Friel, S. (2013). Integrating climate change, food prices and population health. *Food Policy, 43*, 56–66.

Bernstein, A. (2018). *Factors associated with qualifying for abortion funds: An exploratory analysis*.

Briscoe, M. D., et al. (2021). Intersectional indicators: A race and sex-specific analysis of the carbon intensity of well-being in the United States, 1998–2009. *Social Indicators Research, 155*(1), 97–116.

Brisman, A., & South, N. (2019). Green criminology and environmental crimes and harms. *Sociology Compass, 13*(1), e12650.

Buck v. Bell, 274 U.S. 200. (1927).

Burton, E. (2023). Addressing climate change through housing policy: Lessons from the Greater DC Region. *The Washington Post.*

Cafaro, P. (2012). Climate ethics and population policy. *Wiley Interdisciplinary Reviews: Climate Change, 3*(1), 45–61.

Carpio, M. V. (2004). The lost generation: American Indian women and sterilization abuse. *Social Justice, 31*(4 (98)), 40–53.

Chen, S., et al. (2023). Environmental, social, and governance (ESG) performance and financial outcomes: Analyzing the impact of ESG on financial performance. *Journal of Environmental Management, 345*, 118829.

Chijioke, O. C., et al. (2021). Mercantilism in perspective: A historic review. *Global Economy Journal, 21*(01), 2150001.

Chu, A. C., et al. (2022). Agricultural revolution and industrialization. *Journal of Development Economics, 158*, 102887.

Clark, G., O'Rourke, K. H., & Taylor, A. M. (2008). Made in America? The new world, the old, and the industrial revolution. *American Economic Review, 98*(2), 523–528.

Coale, A. J. (1967). The voluntary control of human fertility. *Proceedings of the American Philosophical Society, 111*(3), 164–169.

Colangelo, S. A. (2024). Bridging silos: Environmental and reproductive justice in the climate crisis. *California Law Review, 112*, 1255.

Comack, E. (2018). Corporate colonialism and the "crimes of the powerful" committed against the Indigenous peoples of Canada. *Critical Criminology, 26*(4), 455–471.

Cook-Lynn, E. (1997). Who stole native American studies? *Wicazo Sa Review, 12*(1), 9–28.

Coolsaet, B. (2020). *Environmental justice: Key issues.* Routledge.

CoreLogic. (2023). *Climate change catastrophe report.* Retrieved January 4, 2024, from https://www.corelogic.com/category/intelligence/reports/climate-change/

Coulthard, G. S. (2014). *Red skin, white masks: Rejecting the colonial politics of recognition.*

Crenshaw, K. (1989). Demarginalizing the intersection of race and sex: A Black feminist critique of anti-discrimination doctrine, feminist theory and anti-racist politics. In *The University of Chicago legal forum* (Vol. 140, p. 139).

Crenshaw, K. (1990). Mapping the margins: Intersectionality, identity politics, and violence against women of color. *Stanford Law Review, 43*, 1241.

Crook, M., et al. (2018). Ecocide, genocide, capitalism, and colonialism: Consequences for indigenous peoples and global ecosystems environments. *Theoretical Criminology, 22*(3), 298–317.

Cunningham, M. (2006). Indigenous women's visions of an inclusive feminism. *Development, 49*(1), 55–59.

Daniels, J. (2015, February 16). *The trouble with white feminism: Whiteness, digital feminism, and the intersectional internet. Digital Feminism and the Intersectional Internet.*

Davenport, C. (2024). Deleting climate change from Florida policy. *The New York Times.* Retrieved May 15, 2024, from https://www.nytimes.com/2024/05/15/climate/desantis-climate-change-florida.html

de Allen, G. J. G. (2023). Decolonizing Feminism: Transnational Feminism and Globalization. Margaret McLaren (ed). New York: Rowman & Littlefield, 2017 (ISBN: 978-1-78660-258-9). *Hypatia, 38*(1), e8.

De Beauvoir, S. (1953). *The second sex: Simone de Beauvoir.* Singapore Book.

DDHS. (2025). *Social determinants of health. Health people 2030.* https://odphp.health.gov/healthypeople/priorityareas/social-determinants-health

De Laszlo, H., & Henshaw, P. S. (1954). Plant materials used by primitive peoples to affect fertility. *Science, 119*(3097), 626–631.

de Onís, C. M. (2019). Reproductive justice as environmental justice: Contexts, coalitions, and cautions. In *The Routledge companion to motherhood* (pp. 496–509). Routledge.

Deer, S. (2015). *The beginning and end of rape: Confronting sexual violence in Native America.* U of Minnesota Press.

Deloitte Insights. (2024). *Bridging insurance gaps to prepare homeowners for emerging climate change risks.* Retrieved May 10, 2024, from https://www2.deloitte.com/us/en/insights/industry/financial-services/bridging-the-gap-between-homeowners-insurance-companies-climate-change.html

Desai, Z., & Zhang, Y. (2021). Climate change and women's health: A scoping review. *Geohealth, 5*(9), e2021GH000386.

Dirksen, J., et al. (2022). *Towards economic and social prosperity measurement beyond GDP.*

Dolšak, N., & Prakash, A. (2022). Three faces of climate justice. *Annual Review of Political Science, 25*, 283–301.

Dominelli, L. (2013). Gendering climate change: Implications for debates, policies, and practices. In *Research, action, and policy: Addressing the gendered impacts of climate change* (pp. 77–93). Springer.

Dworkin, G. (1970). Acting freely. *Noûs*, 367–383.

Edmans, A. (2021). Grow the pie. *Journal of the History of Economic Thought, 16*, 592.

Edmans, A. (2024, January 20). *Rational sustainability.* London Business School.

Eisenstadt v. Baird, 405 U.S. 438 (1972).

Enarson, E. P. (2012). *Women confronting natural disaster: From vulnerability to resilience* (p. 245). Lynne Rienner Publishers.

Ergas, C., & York, R. (2012). Women's status and carbon dioxide emissions: A quantitative cross-national analysis. *Social Science Research, 41*(4), 965–976.

European Union. (2023). *Corporate sustainability reporting*. Retrieved April 10, 2024, from https://finance.ec.europa.eu/capital-markets-union-and-financial-markets/company-reporting-and-auditing/company-reporting/corporate-sustainability-reporting_en

Everett, B. G., & Taylor, C. J. (2024). Abortion and women's future socioeconomic attainment. *American Sociological Review, 89*(6), 1044–1074.

Ewall, M. (2012). Legal tools for environmental equity vs. environmental justice. *Sustainable Development Law & Policy, 13*, 4.

Federal Reserve Economic Data. (2024). *Economic data*. Retrieved May 1, 2024, from https://fred.stlouisfed.org

Flitter, A. (2023, July 14). Insurer's retreat in Florida signals crisis with no easy fix. *The New York Times*.

Forbes. (2024). 5 of the best oil and gas stocks to buy for high crude prices. *Forbes*. Retrieved May 16, 2024, from https://www.forbes.com/sites/investor-hub/article/best-oil-gas-stocks-amid-high-crude-oil-prices/?sh=7b4ca6af2b73

Foster, D. G., et al. (2022). Socioeconomic outcomes of women who receive and women who are denied wanted abortions in the United States. *American journal of public health, 112*(9), 1290–1296.

Frankfurt, H. (1971). Freedom of the will and the concept of a person. *The Journal of Philosophy, 68*(5), 6–10.

Fremstad, A., et al. (2019). Work hours and CO2 emissions: Evidence from US households. *Review of Political Economy, 31*(1), 42–59.

Friedman, M. (1970, September 13). *The social responsibility of business is to increase its profits* (pp. 122–126). New York Times Magazine.

Fullerton, J. (2010, June 28). *Is sustainable capitalism an oxymoron?* Capital Institute.

Fullerton, J. (2015). *Regenerative capitalism* (pp. 1–120). Capital Institute.

Gallup Survey. (2018). *Global warming age gap: Younger Americans most worried*. Gallup.

Gallup Survey. (2023). *Americans' preference for larger families highest since 1971*. Gallup.

Garrett, D., & Ivanov, I. (2024). *Gas, guns, and governments: Financial costs of anti-ESG policies*. Jacobs Levy Equity Management Center for Quantitative Financial Research Paper.

Gerrard, M. B. (2017). Climate change and human trafficking after the Paris agreement. *University of Miami Law Review, 72*, 345.

Giddens, A. (1984). *The constitution of society.*

Giddens, A. (1990). *The consequences of modernity.*

Gilio-Whitaker, D. (2019). *As long as grass grows: The Indigenous fight for environmental justice, from colonization to Standing Rock.* Beacon Press.

Goldsmith, E. (1997). Development as colonialism. *The Ecologist, 27*(2), 69–77.

Gonzalez, C. G. (2021). Racial capitalism, climate justice, and climate displacement. In *Oñati socio-legal series, symposium on climate justice in the Anthropocene* (Vol. 11, pp. 108–147).

Gore, A., & Blood, D. (2010). Toward sustainable capitalism. *The Wall Street Journal.*

Grann, D. (2017). *Killers of the flower moon: The Osage murders and the birth of the FBI.* Vintage.

Griswold v. Connecticut, 381 U.S. 479 (1965).

Guzzo, K. B., & Hayford, S. R. (2023). Evolving fertility goals and behaviors in current US childbearing cohorts. *Population and Development Review, 49*(1), 7–42.

Hammond, G. P. (2006). People, planet and prosperity: The determinants of humanity's environmental footprint. In *Natural resources forum* (Vol. 30, pp. 27–36). Blackwell Publishing Ltd.

Hannusch, A., et al. (2022). *The economics of fertility: A new era (no. 17212).* CEPR discussion papers.

Hardin, G. (1968). The tragedy of the commons: The population problem has no technical solution; it requires a fundamental extension in morality. *Science, 162*(3859), 1243–1248.

Harris Poll. (2022). *Personal independence behind declining birth rates.* The Harris Poll.

Harris Poll. (2023). *The state of American capitalism.* The Harris Poll.

Harris v. McRae, 448 U.S. 297 (1980).

Helbing, D., & Helbing, D. (2015). *What the digital revolution means for us* (pp. 177–187). Springer International Publishing.

Henrici, J., Childers, R., & Shaw, E. (2015). *Get to the bricks: The experiences of black women from New Orleans public housing after hurricane Katrina.* Institute for Women's Policy Research.

Hickman, C., et al. (2021). Climate anxiety in children and young people and their beliefs about government responses to climate change: A global survey. *The Lancet Planetary Health, 5*(12), e863–e873.

Hooks, B. (2000). *Feminism is for everybody: Passionate politics.* Pluto Press.

Howard, P. (2023). *Home insurance prices up 21% as homeowners are left to deal with climate change, turbulent market.* 2023 Policygenius Home Insurance Pricing Report. Policygenius.

Ip, G., & Adamy, J. (2024). Suddenly there aren't enough babies. The whole world is alarmed. *The Wall Street Journal*. Retrieved May 14, 2024, from https://www.wsj.com/world/birthrates-global-decline-cause-ddaf8be2

IWPR. (2010). Women, disasters, and hurricane katrina. institute for women's policy research. https://iwpr.org/wp-content/uploads/2020/09/D492.pdf

Jackson, T. (2016). *Prosperity without growth: Foundations for the economy of tomorrow*. Taylor & Francis.

Jägermeyr, J., et al. (2021). Climate impacts on global agriculture emerge earlier in new generation of climate and crop models. *Nature Food, 2*(11), 873–885.

Johnson-Hanks, J. A., et al. (2011). *Understanding family change and variation: Toward a theory of conjunctural action* (Vol. 5). Springer Science & Business Media.

Kalkandelen, K., & O'Byrne, D. (2017). On ecocide: Toward a conceptual framework. *Distinktion: Journal of Social Theory, 18*(3), 333–349.

Kant, I. (1785). *Groundwork of the metaphysics of morals*.

Kearney, M. S., et al. (2022). The puzzle of falling US birth rates since the Great Recession. *Journal of Economic Perspectives, 36*(1), 151–176.

Keerthiratne, S., & Tol, R. S. (2018). Impact of natural disasters on income inequality in Sri Lanka. *World Development, 105*, 217–230.

Kelly, M. (2023). *Wealth supremacy: How the extractive economy and the biased rules of capitalism drive today's crises*. Berrett-Koehler Publishers.

Keynes, J. M. (1933). The means to prosperity. In *Essays in persuasion* (pp. 335–366). Palgrave Macmillan UK.

Keynes, J. M. (1936). *General theory. Reprint*. Harbinger.

Khan, I., et al. (2023). Striving for carbon neutrality and economic prosperity in the top ten emitting countries: Testing N shape Kuznets curve hypothesis. *Journal of Cleaner Production, 429*, 139641.

Klepp, S. (2017). Climate change and migration. In *Oxford research encyclopedia of climate science*.

Lane, R., & McNaught, R. (2009). Building gendered approaches to adaptation in the Pacific. *Gender & Development, 17*(1), 67–80.

Larocque, M. (1993). Ecofeminism: Women, animals, nature. *Canadian Woman Studies/les cahiers de la femme*.

Lazarus, E. (1999). *Black hills white justice: The Sioux nation versus the United States, 1775 to the present*. U of Nebraska Press.

Lee, Y. S. (2023). Shareholder primacy as an untenable corporate norm. *Annals of Corporate Governance, 8*(1), 1–50.

Lee, H., et al. (2023). *IPCC, 2023: Climate change 2023: Synthesis report, summary for policymakers*. Contribution of Working Groups I, II and III to the Sixth Assessment Report of the Intergovernmental Panel on Climate Change [Core Writing Team, H. Lee and J. Romero (eds.)]. IPCC.

Leiserowitz, A., et al. (2020a). *Climate change in the American mind: November 2019.*

Leiserowitz, A., et al. (2020b, April). *Politics and global warming.*

Licata, J. C. (2020). *The politics of Norplant: Feminism, civil rights, and social policy in the 1990s.* The University of North Carolina at Greensboro.

Lima de Miranda, K., & Snower, D. J. (2020). Recoupling economic and social prosperity. *Global Perspectives, 1*(1).

Lumsden, S. (2016). Reproductive justice, sovereignty, and incarceration: Prison abolition politics and California Indians. *American Indian Culture and Research Journal, 40*(1), 33–46.

Lux, M. K. (2016). *Separate beds: a history of Indian hospitals in Canada, 1920s–1980s.* University of Toronto Press.

Lynch, M. J. (1990). The greening of criminology: A perspective for the 1990s. Reprinted in N. South & P. Beirne (Eds.). In *Green criminology.*

Ma, Y., et al. (2023). From traditional innovation to green innovation: How an occurrence of natural disasters influences sustainable development? *Sustainable Development.*

MacDonald, R. (2005). How women were affected by the tsunami: A perspective from Oxfam. *PLoS Medicine, 2*(6), e178.

MacGregor, S. (2017). Moving beyond impacts: More answers to the 'gender and climate change's question. In *Understanding climate change through gender relations* (pp. 15–30). Routledge.

Mackenzie, C. (2021). Relational equality and the debate between externalist and internalist theories of relational autonomy. In *Autonomy and equality* (pp. 32–56). Routledge.

Mal, S., et al. (2018). Introducing linkages between climate change, extreme events, and disaster risk reduction. In *Climate change, extreme events and disaster risk reduction: Towards sustainable development goals* (pp. 1–14).

Marsh, M., & Ronner, W. (2008). *The fertility doctor: John Rock and the reproductive revolution.* JHU Press.

Martin, J. W. (2018). *Banana cowboys: The United Fruit Company and the culture of corporate colonialism.* University of New Mexico Press.

Martínez, D. (2013). Remembering the thirty-eight: Abraham Lincoln, the Dakota, and the US war on barbarism. *Wicazo Sa Review, 28*(2), 5–29.

Martinez, G. M., & Daniels, K. (2023). Fertility of men and women aged 15–49 in the United States: National survey of family growth, 2015–2019. *National Health Statistics Reports,* (179), 1–22.

Martinez-Diaz, L., & Keenan, J. M. (Eds.). (2020). *Managing climate risk in the US financial system.* US Commodity Futures Trading Commission.

Marx, K. (1848). The communist manifesto. In *Ideals and ideologies* (pp. 243–255). Routledge.

Marx, K. (1907). *Capital: The process of circulation of capital*, ed. by Frederick Engels, tr. from the 2d German ed. by Ernest Untermann (Vol. 2). CH Kerr.

Mayer, C. (2021). The future of the corporation and the economics of purpose. *Journal of Management Studies, 58*(3), 887–901.

Mazurek, J. (2011). Environmental Kuznets curve: A tie between environmental quality and economic prosperity. *Economics and Management.*

McCauley, D., & Heffron, R. (2018). Just transition: Integrating climate, energy, and environmental justice. *Energy Policy, 119*, 1–7.

McDermott-Levy, R., et al. (2021). Factors that influence climate change-related mortality in the United States: An integrative review. *International Journal of Environmental Research and Public Health, 18*(15), 8220.

McGee, J. A., et al. (2020). Gender inequality, reproductive justice, and decoupling economic growth and emissions: A panel analysis of the moderating association of gender equality on the relationship between economic growth and CO2 emissions. *Environmental Sociology, 6*(3), 254–267.

McQuillan, J., et al. (2015). The importance of motherhood and fertility intentions among US women. *Sociological Perspectives, 58*(1), 20–35.

Megura, M., & Gunderson, R. (2022). Better poison is the cure? Critically examining fossil fuel companies, climate change framing, and corporate sustainability reports. *Energy Research & Social Science, 85*, 102388.

Mihesuah, D. A. (2000). A few cautions at the millennium on the merging of feminist studies with American Indian Women's Studies. *Signs: Journal of Women in Culture and Society, 25*(4), 1247–1251.

Mill, J. S. (1974). *On liberty (1859).*

Million, D. (2009). Felt theory: An Indigenous feminist approach to affect and history. *Wicazo Sa Review, 24*(2), 53–76.

Mohanty, C. T. (2003). "Under western eyes" revisited: Feminist solidarity through anticapitalist struggles. *Signs: Journal of Women in culture and Society, 28*(2), 499–535.

Mooney, J. (1991). *The ghost-dance religion and the Sioux outbreak of 1890* (Vol. 176). U of Nebraska Press.

Morgan, S. P., & Rackin, H. (2010). The correspondence between fertility intentions and behavior in the United States. *Population and Development Review, 36*(1), 91–118.

Morgensen, S. L. (2010). Settler homonationalism: Theorizing settler colonialism within queer modernities. *GLQ: A Journal of Lesbian and Gay Studies, 16*(1-2), 105–131.

Morning Consult. (2018). To breed or not to breed? *The New York Times.*

Murtaugh, P. A., & Schlax, M. G. (2009). Reproduction and the carbon legacies of individuals. *Global Environmental Change, 19*(1), 14–20.

Nandi, A., Mazumdar, S., & Behrman, J. R. (2018). The effect of natural disaster on fertility, birth spacing, and child sex ratio: Evidence from a major earthquake in India. *Journal of Population Economics, 31*, 267–293.

Nash, J. (1953). Two-person cooperative games. *Econometrica: Journal of the Econometric Society*, 128–140.

National Archives. (2016). *Indian Census roles, 1885–1940.* Legal and Administrative Background: The U.S. National Archives and Records Administration.

Nedelsky, J. (1989). Reconceiving autonomy: Sources, thoughts and possibilities. *Yale Journal of Law and Feminism, 1*, 7.

Neumayer, E., & Plümper, T. (2007). The gendered nature of natural disasters: The impact of catastrophic events on the gender gap in life expectancy, 1981–2002. *Annals of the Association of American Geographers, 97*(3), 551–566.

Nietzsche, F. (1881). *Nietzsche: Daybreak: Thoughts on the prejudices of morality.* Cambridge University Press.

Nietzsche, F. (2008). *Thus spoke Zarathustra: A book for everyone and nobody.* Oxford University Press.

NORC. (2023). *National opinion on the state of the nation's economy.* National Opinion Research Center. University of Chicago.

Nussbaum, M. C. (2011). *Creating capabilities: The human development approach.*

OECD. (2024). *Society at a glance 2024: OECD social indicators.* OECD Publishing. https://doi.org/10.1787/918d8db3-en

Office of the Comptroller of the Currency. (2021). *OCC finalizes rule requiring large banks to provide fair access to bank services, capital, and credit.* Retrieved April 14, 2024, from https://www.occ.gov/news-issuances/news-releases/2021/nr-occ-2021-8.html

Office of the Governor of the State of Hawaii. (2023). Seventh Proclamation Relating to Wildfires. https://governor.hawaii.gov/wp-content/uploads/2023/09/2309064.pdf

Pachauri, R. K., & Meyer, L. A. Fifth assessment report of the intergovernmental panel on climate change, ed. RK Pachauri and LA Meyer (Geneva, Switzerland: IPCC, 2014), 1–151.

Pan, L., & Xu, J. (2012). Housing price and fertility rate. *China Economic Journal, 5*(2–3), 97–111.

Panda, R., & Maity, M. (2021). Global warming and climate change on earth: Duties and challenges of human beings. *International Journal of Research in Engineering, Science and Management, 4*(1), 122–125.

Paterson, J., et al. (2014). Health care facilities resilient to climate change impacts. *International Journal of Environmental Research and Public Health, 11*(12), 13097–13116.

Patten, S. N. (1902). *The theory of prosperity.* Macmillan.

Patten, S. N. (1968). *The new basis of civilization* (Vol. 1). Harvard University Press.

Pearse, R. (2017). Gender and climate change. *Wiley Interdisciplinary Reviews: Climate Change, 8*(2), e451.

Pegoraro, L. (2015). Second-rate victims: The forced sterilization of Indigenous peoples in the USA and Canada. *Settler Colonial Studies, 5*(2), 161–173.

Peretto, P. F. (2021). Through scarcity to prosperity: Toward a theory of sustainable growth. *Journal of Monetary Economics, 117*, 243–257.

Pew Research Center. (2014). *Climate change: Key data points from Pew Research.* Pew Research Center.

Pew Survey. (2021). *Growing share of childless adults in U.S. don't expect to ever have children.* Pew Research Center.

Pew Survey. (2024). *Americans' top policy priority for 2024: Strengthening the economy.* Pew Research Center.

Planned Parenthood v. Casey, 505 U.S. 833 (1992).

Prucha, F. P. (1995). *The great father: The United States government and the American Indians* (Vol. 1). U of Nebraska Press.

Pumphrey, S. R. (2014). *Reclaiming my body: Black women and the fight for reproductive justice.*

Quijano, A. (2000). Coloniality of power and Eurocentrism in Latin America. *International Sociology, 15*(2), 215–232.

Qureshi, M. A., et al. (2021). Do ESG endeavors assist firms in achieving superior financial performance? A case of 100 best corporate citizens. *SAGE Open, 11*(2), 21582440211021598.

Rabbitt, M. P., et al. (2023). *Household food security in the United States in 2022.*

Raju, E., et al. (2022). Stop blaming the climate for disasters. *Communications Earth & Environment, 3*(1), 1.

Raworth, K. (2017). *Doughnut economics: Seven ways to think like a 21st-century economist.* Chelsea Green Publishing.

Renzetti, C. M. (2018). Feminist perspectives. In *Routledge handbook of critical criminology* (pp. 74–82). Routledge.

Retsikas, K., & Marsden, M. (2018). Alternate modes of prosperity. *HAU: Journal of Ethnographic Theory, 8*(3), 596–609.

Rindfuss, R. R., Bumpass, L., & St. John, C. (1980). Education and fertility: Implications for the roles women occupy. *American Sociological Review*, 431–447.

Roberts, D. E. (1999). *Killing the black body: Race, reproduction, and the meaning of liberty.* Vintage.

Robinson, M. (2019). *Climate justice: A man-made problem with a feminist solution.* Bloomsbury Publishing.

Rockström, J., et al. (2009). A safe operating space for humanity. *Nature, 461*(7263), 472–475.

Rodrik, D., & Stantcheva, S. (2021). *A policy matrix for inclusive prosperity (No. w28736).* National Bureau of Economic Research.

Roe v. Wade, 410 U.S. 113 (1973).

Ross, L. (Ed.). (2016). *Settler colonialism and the legislating of criminality.* American Indian Studies Center, University of California.

Ross, L. J. (2017). Conceptualizing reproductive justice theory: A manifesto for activism. In *Radical reproductive justice: Foundations, theory, practice, critique* (pp. 170–232).

Ross, L., & Solinger, R. (2017). *Reproductive justice.* University of California Press.

Ross, L., et al. (2016). *Undivided rights: Women of color organizing for reproductive justice.* Haymarket Books.

Rust v. Sullivan, 500 U.S. 173 (1991).

Salm, L., et al. (2021). How climate change interacts with inequity to affect nutrition. *Wiley Interdisciplinary Reviews: Climate Change, 12*(2), e696.

Sasser, J. S. (2018). *On infertile ground: Population control and women's rights in the era of climate change.* NYU Press.

Sasser, J. S. (2024). At the intersection of climate justice and reproductive justice. Wiley Interdisciplinary Reviews: Climate Change, 15(1), e860.

Schoen, R., et al. (1997). Why do Americans want children? *Population and Development Review*, 333–358.

Securities and Exchange Commission. (2024). *The enhancement and standardization of climate-related disclosures for investors.* Retrieved April 10, 2024, from https://www.sec.gov/files/rules/final/2024/33-11275.pdf

Sen, A. (1979). *Equality of what?* (Vol. 1).

Sewell, W. H., Jr. (1992). A theory of structure: Duality, agency, and transformation. *American Journal of Sociology, 98*(1), 1–29.

Sewell Jr, W. H., & Sewell, W. H. (2005). *Logics of history: Social theory and social transformation.* University of Chicago Press.

Sharp, H. (2021). *Spinoza and feminism. A companion to Spinoza* (pp. 422–430).

Sheu, J. C., et al. (2021). Potential impact of climate change on human trafficking: A narrative review. *The Journal of Nervous and Mental Disease, 209*(5), 324–329.

Silko, L. M. (2013). *Yellow woman and a beauty of the spirit.* Simon and Schuster.

Simpson, A. (2011). Settlement's secret. *Cultural Anthropology, 26*(2), 205–217.

Simpson, A. (2016). The state is a man: Theresa Spence, Loretta Saunders, and the gender of settler sovereignty. *Theory & Event, 19*(4).

Simpson, L. B. (2017). *As we have always done: Indigenous freedom through radical resistance.* U of Minnesota Press.

Smith, A. (1776). *An inquiry into the nature and causes of the wealth of nations: Volume One.* Printed for W. Strahan; and T. Cadell.

Sobotka, T., et al. (2011). Economic recession and fertility in the developed world. *Population and Development Review, 37*(2), 267–306.

Sorensen, C., et al. (2018). Climate change and women's health: Impacts and policy directions. *PLoS Medicine, 15*(7), e1002603.

Speth, J. G. (2009). From crisis to sustainability. In L. Mazur (Ed.), *A pivotal moment: Population, justice, and the environmental challenge* (pp. 329–338). Island Press.

Stevenson, W. (1999). Colonialism and first nations women in Canada. In *Scratching the surface: Canadian anti-racist feminist thought* (pp. 49–80).

Stone, L. (2018). *Declining fertility in America*. American Enterprise Institute.

Sullivan & Cromwell LLP. (2024). *States require "Fair Access" to financial services.* Retrieved May 1, 2024, from https://www.sullcrom.com/SullivanCromwell/_Assets/PDFs/Memos/States-Require-Fair-Access-Financial-Services.pdf

Tacoli, C., et al. (2015). *Urbanization, rural-urban migration, and urban poverty.* Human Settlements Group, International Institute for Environment and Development.

Tallbear, K. (2012). *The US-Dakota War of 1862: Ethnic strife or the political economy of war?* Oak Lake Writers Society.

TallBear, K. (2016). Annual meeting: The US-Dakota War and failed settler kinship. *Anthropology News, 57*(9), e92–e95.

Teran, J. (2016). The violent legacies of the California missions: Mapping the origins of Native women's mass incarceration. *American Indian Culture and Research Journal, 40*(1), 19–32.

Tesla, B., et al. (2018). Temperature drives Zika virus transmission: Evidence from empirical and mathematical models. *Proceedings of the Royal Society B, 285*(1884), 20180795.

Texas Association of Business Chambers of Commerce Foundation. (2024). *The potential economic and tax revenue impact of Texas' fair access laws.* Retrieved March 13, 2024, from https://cb9cdd3c-61f1-494f-94da-c77c057de62c.usrfiles.com/ugd/cb9cdd_d0b8b35ea13b4294be456e4113abef3b.pdf

Thornton, R. (1987). *American Indian holocaust and survival: A population history since 1492* (Vol. 186). University of Oklahoma Press.

Trinitapoli, J., & Yeatman, S. (2018). The flexibility of fertility preferences in a context of uncertainty. *Population and Development Review, 44*(1), 87.

Tuck, E., & Yang, K. W. (2021). Decolonization is not a metaphor. *Tabula Rasa*, (38), 61–111.

Turnbull, S. (2011). Achieving environmentally sustainable prosperity. In *Toward a good society in the twenty-first century: Principles and policies* (pp. 75–103). Palgrave Macmillan US.

United Nations. (2024). *Global Resources Outlook 2024: Bend the trend-Pathways to a liveable planet as resource use spikes.*

U. N. Women (2025). The gender snapshot 2025. https://www.unwomen.org/en/digital-library/publications/2025/09/progress-on-the-sustainable-development-goals-the-gender-snapshot-2025

U.S. Department of Agriculture, Economic Research Service. (n.d.-a). *Definitions of food security.* Retrieved January 4, 2024, from https://www.ers.usda.gov/topics/food-nutrition-assistance/food-security-in-the-u-s/definitions-of-food-security/

U.S. Department of Agriculture, Economic Research Service. (n.d.-b). *Key statistics & graphics.* Retrieved January 4, 2024, from https://www.ers.usda.gov/topics/food-nutrition-assistance/food-security-in-the-u-s/key-statistics-graphics/

U.S. Department of Health and Human Services. (n.d.). *Healthy people 2030.* Retrieved January 4, 2024, from https://health.gov/healthypeople/priority-areas/social-determinants-health/literature-summaries/food-insecurity

United Nations Population Fund (UNFPA). (2025). *State of World Population 2025: The Real Fertility Crisis-The Pursuit of Reproductive Agency in a Changing World.* Stylus Publishing, LLC.

United States Department of the Treasury. (2023). *The impact of climate change on American household finances.*

USDA. (2025). *Definitions of food security. Food Security in the U.S.* https://www.ers.usda.gov/topics/foodnutrition-assistance/food-security-in-the-us/definitions-of-food-security.

Venkatraman, R. (2020). *A new vision for advancing our movement for reproductive health, reproductive rights, and reproductive justice (2005), by Asian communities for reproductive justice (ACRJ).* Embryo Project Encyclopedia.

Wackernagel, M., & Rees, W. (1998). *Our ecological footprint: Reducing human impact on the earth* (Vol. 9). New Society Publishers.

Wang, J. (2010). Food security, food prices and climate change in China: A dynamic panel data analysis. *Agriculture and Agricultural Science Procedia, 1*, 321–324.

Wang, Z. X., & Li, Q. (2019). Modelling the nonlinear relationship between CO2 emissions and economic growth using a PSO algorithm based grey Verhulst model. *Journal of Cleaner Production, 207*, 214–224.

Washington Post. (2023). *The human limit.* Retrieved January 3, 2024, from https://washingtonpost.com/climate-environment/climate-lab/

Weber, M. (1904). *The protestant ethic and the spirit of capitalism.*

Whittenbury, K. (2013). Climate change, women's health, wellbeing, and experiences of gender-based violence in Australia. In *Research, action, and policy: Addressing the gendered impacts of climate change* (pp. 207–221). Springer.

Wicks, E. (2016). *The state and the body: Legal regulation of bodily autonomy.* Bloomsbury Publishing.

Wilkinson, R. G., Marmot, M., & World Health Organization. (1998). *The solid facts: Social determinants of health (No. EUR/ICP/CHVD 03 09 01).* WHO Regional Office for Europe.

World Economic Forum. (2024). *The global risks report 2024 19th edition*. Cologny, Switzerland.

World Health Organization. (2008). *Social determinants of health (No. SEA-HE-190)*. WHO Regional Office for South-East Asia.

Wu, J., et al. (2018). Economic development and declining vulnerability to climate-related disasters in China. *Environmental Research Letters, 13*(3), 034013.

Wynes, S., & Nicholas, K. A. (2017). The climate mitigation gap: Education and government recommendations miss the most effective individual actions. *Environmental Research Letters, 12*(7), 074024.

Xie, W., Rose, A., Li, S., He, J., Li, N., & Ali, T. (2018). Dynamic economic resilience and economic recovery from disasters: A quantitative assessment. *Risk Analysis, 38*(6), 1306–1318.

Yoshino, K., & Glasgow, D. (2024). DEI is under attack. Here's how companies can mitigate the legal risks. *Harvard Business Review*. Retrieved April 4, 2024, from https://hbr.org

Young, I. (2011). *Justice and the politics of difference*. Princeton University Press.

Zein, L. F., & Setiawan, A. R. (2017). *General overview of ecofeminism*. OSF Preprints.

Zottarelli, L. K. (2008). Post-hurricane Katrina employment recovery: The interaction of race and place. *Social Science Quarterly, 89*(3), 592–607.

Index

S. Haley Knowles, *Diminished Prosperity*,
https://doi.org/10.1007/978-3-032-07068-5

W

9783032070678